RELATIVELY
SPACIOUS

Published in the UK in 2021 by Uniity Publishing

Copyright © Gary Bridges 2021

Gary Bridges has asserted their right under the Copyright, Designs and Patents Act, 1988, to be identified as the author of this work.

All rights reserved. No part of this book may be reproduced, stored in a retrieved system or transmitted, in any form or by any means, electronic, mechanical, scanning, photocopying, recording or otherwise, without the prior permission of the author and publisher.

Paperback ISBN 978-1-7399074-0-2
eBook ISBN 978-1-7399074-1-9

Cover design and typeset by SpiffingCovers.com

RELATIVELY
SPACIOUS

SPACE AND TIME AND OUR PLACE WITHIN IT

GARY BRIDGES

For my lovely dad who is sadly no longer with us.

To my beautiful daughter, Madelaine, who is surely a gift from the stars.

To Nicholas and Mathew, how proud I am of the lovely men you have become despite the challenges in your early years.

To Alex, my youngest child, you have the makings of greatness; all you need to do is believe.

CONTENTS

A little Introduction   1

CHAPTER ONE
The Power of the Sun   11

CHAPTER TWO
Of Gravity, Mass and the Speed of Light   24

CHAPTER THREE
In the Beginning   39

CHAPTER FOUR
Of Planets and the Solar System   50

CHAPTER FIVE
Amazing Earth   71

CHAPTER SIX
A Bit about Moons   81

CHAPTER SEVEN
Awesome Space Technology   94

CHAPTER EIGHT
Life   118

CHAPTER NINE
Alien Civilisations   135

A little Introduction

I guess we have all contemplated our existence at some point. Who are we? Where did we come from? Where did we really come from? What does it actually mean to be alive?

Are we beginning to understand the miracle of living?

I guess you could call life a miracle, but we certainly don't understand it.

The word miracle is defined in the English dictionary as:

> *An extraordinary and welcome event that is not explicable by natural or scientific laws and is therefore attributed to a divine agency.* **(Oxford Dictionaries)**

Until quite recently, this is what most people attributed to things they didn't understand.

In fact, when you come to understand a little of the universe and how it, and we, came to be, it does seem like there was a divine blueprint for our existence. Fortunate event followed fortunate event that has allowed for you and me to be here now, to exist.

Kind of like a million double sixes in a row on a celestial dice roll.

There is a vast amount of information that scientists have been able to gather from the stars and galaxies, but there is also an astronomical amount we don't know.

Incredible amounts of information come pouring into scientists' databases constantly from large communication dishes, satellites, telescopes, space telescopes, interplanetary probes, surface rovers on Mars and, not to mention, the International Space Station. There are still so many unanswered questions though.

What I am hoping to share with you in my book is a bit of both; some mind-boggling facts along with some interesting theories about some of the mysteries of the universe.

I had always wanted to write a book; it was going to be science fiction though, about a large, ark-type spacecraft drifting through space and its encounters with black holes and neutron stars, discovering planets where life was just starting out or planets where life was barely hanging on.

So, what changed?

Well, to be honest, I don't really know. I think, as I began to read more and more science books, I realised that the universe we live in is a much more strange and incredible place than anything I ever could have imagined.

As I am sure you can imagine it wasn't always straightforward and more than once I had all but given up. Luckily there were times that spurred me on to write or jolted me back to the unfinished manuscript. Whilst I was away abroad with some

A little Introduction

friends I found myself sitting on the terrace of a bar while the sun was setting. We were all a little transfixed by the beautiful scene. I noticed a few of my group were very red in the face. They had been careless, spending time out of the shade without adequate protection from the colossal 864,000-mile-wide nuclear reactor we now found ourselves staring at.

The sun was magnificent that evening, an immense, deep-singed, orange semi sphere, balancing, half submerged on the Mediterranean horizon.

I casually asked one of my friends how far away he thought the sun might be. He looked uncomfortable at first, then after a few seconds blurted out, "Ten thousand miles."

I smiled thinking he was joking. He wasn't.

I was staggered. Upon more questions, like "Do you know how far away the moon is?" or "Do you know what gravity is?", he seemed to become hot and agitated and eventually, after pulling some strange faces, jumped up from his seat and walked off, leaving his drink unfinished!

I was still quite shocked; this friend wasn't stupid, in fact he had the better education of the two of us and was a very successful businessman. Somehow, though, he had very limited knowledge of the world around him and clearly didn't want to appear ignorant about it.

I began to realise this was quite a common thing. People were happy to go about their daily routines without knowing about the really big stuff, the really interesting stuff, the stuff that explains our very existence and ultimately our fate, not just as a species but as a planet, as a solar system. Space isn't somewhere else, a place beyond human boundaries or comprehension, space is all around us. Travel upwards for a hundred miles and you're there. You would travel further for a holiday abroad.

At an early age, the solar system had caught my attention. I became kind of hooked on the planets and their movements around the sun. I began to get an understanding of gravity and planetary orbits.

Which is all rather strange really, as I wasn't particularly good at maths, or anything else, come to think of it.

I struggled to get dirty washing into the laundry bin or remember things on a shopping errand, even if I was handed a list. I was a poor student, easily distracted by a runaway imagination. I found myself constantly in trouble, much to my parents' despair!

Growing up, I was lucky to have a good library though. Here, I spent days looking through all the science books I could find. On bright, summer days, the light of the sun would illuminate a little corner of the quiet room through an arched window, and I would sit for hours on a rather unstable, wonky wooden chair reading, fascinated. On winter days, it was a haven from the cold and an excuse to get stuck into even more books.

The problem I had, though, was that I struggled to understand most of them; a great deal of them were, well, just beyond me.

A little Introduction

Take a look at the paragraph below from Neil deGrasse Tyson about the universe, a trillionth of a second from its conception, and you might see what I mean:

> All the while, the interplay of matter in the form of subatomic particles, and energy in the form of photons (massless vessels of light energy that are as much waves as they are particles) was incessant.

I really wanted to know what these words meant. These science books I was attempting to read were written for people who must have had a much greater understanding than me. I wanted to read a book that would explain these things without having to have a degree in chemistry or astrophysics.

There must be a book, a kind of Rosetta Stone for space science, so everybody could enjoy the wonders of space, the planets and the universe.

There was no book, or maybe there was, just not in my little library though. This was, of course, well before the time of Amazon or the internet.

So, after a while I realised, I was going to have to start at the very basics and teach myself from there. And that's how it started, with atoms, I mean.

If I was going to be able to understand the big stuff, I was going to have to understand the very little stuff first. I needed to understand light and chemicals and elements. I was going to have to find out what an atom was and where I could find one. I became curious about what elements made up these enormous bodies in space that I had become so captivated with.

The more I read, the more I needed to know. The problem was an answer to one question almost always opened up more questions.

I would look at pictures in books and it would seem like questions were falling from the pages.

Why are all the planets nearly perfect spheres? Not most of them but all of them. Why aren't they triangular or just totally irregular? What process could make that happen?

How does our sun continue to shine for billions of years? Surely it would have used up all its fuel by now?

Why is gravity so weak that we can leap into the air but strong enough to enable a moon, hundreds of thousands of miles away, to affect our oceans?

How fast does the earth travel through the void of space?

What is a black hole? And why is there a black hole?

How did the universe begin?

I am not an astronomer or a mathematician or even a writer, as it happens. I am just attempting to write a book to explain some of the mysteries, and amazing facts and theories, about the universe that I think people would like to know.

A little Introduction

Guys like Brian Cox and Bill Bryson have done a fantastic job of doing this, and I highly recommend reading their works.

They have both been a real inspiration to me.

However, I'm still surprised at the lack of knowledge a lot of people have when it comes to these subjects. Most people I asked couldn't tell me the nearest or furthest planet from the sun, or how far away they are.

I have learnt, through reading and studying, that science is amazing.

We don't feel the need now to sacrifice a small child when a volcano erupts or worship a bright light in the sky because we have no other explanation than a divine one.

Science searches for the answers, for the truth.

Science had a pretty tough time getting going though, but when it did, the leaps forward have been incredible. So, what held science back for so long?

The great German physicist Max Planck once wrote:

> A new scientific truth does not triumph by convincing its opponents and making them see the light, but rather because its opponents eventually die, and a new generation grows up that is familiar with it.

Basically, I believe he is saying scientists can be stubborn and closed to radical thinking, so new ideas can only prevail upon the death of the scientists who object to the idea. This, I am sure, must have been the case. However, some greatly revered thinkers have possibly halted progress even after death.

The great Aristotle was thought of so highly that when he stated something, it was generally considered fact.

He strongly advocated that the sun and the planets revolved around the earth, and thus it became almost fact for nearly two thousand years.

In 1616, when Galileo announced it was the planets that revolved around the sun, he was accused of heresy, and threatened with torture and death by the Catholic Church, until he eventually and reluctantly withdrew his work. He was still to spend the remainder of his life under house arrest though.

When Charles Darwin was studying animals and their environments in South America and the Galapagos Islands on his journey upon the *Beagle* in 1831, had thoughts of natural selection been evolving in his mind? Did he worry what the Church would think of his theories? If he didn't, then he certainly did later in his life. He withheld his works from the world due to this concern and only published them when he became aware that Alfred Russel Wallace was thinking along the same lines. Worried that his life's work was about to be hijacked, he was forced into publication.

Darwin's *On the Origin of Species by Means of Natural Selection* is still a hotbed of discussion between science and religion even today.

A little Introduction

Society just didn't seem to like, or indeed want, change; for better or worse.

At some point, though, people became more willing to accept new ideas, possibly helped by Einstein's theory of relativity and other incredible scientific discoveries in the early 20th century.

Individuals, without the threat of torture or death, seemed to become more resilient to ridicule and criticism.

Orville and Wilbur Wright were bicycle shop owners from Ohio that had an incredible desire; to create a flying machine out of lightweight timber and fabric.

This was considered ridiculous by most people. In fact, the *Washington Post* declared: "It is a fact that man cannot fly" after witnessing a failed attempt by the brothers.

And yet, on the 17th of December 1903 at Kitty Hawk, North Carolina, they did fly. And within 70 years of this, human beings had set foot on another celestial body in space.

Quite an achievement, when you consider the developments over the last three thousand years.

Marconi, Curie, Edison and Pasteur to name but a few, all endured similar criticism or scepticism but came through it, probably helped by the fact they knew their head would remain attached to their body, despite what people thought.

What would these sceptics think of today's accomplishments? What would they think about the fact that there is more technology in your electronic car key fob than there was used to land a man on the moon? What about landing a spacecraft on an asteroid travelling at 63,000 miles an hour? Or, how about remotely manoeuvring a rover on another planet millions of miles away?

This is a book about space science facts and theories. A book where we can learn that space is not what you think it is, time as we know it is an illusion, and to get anywhere in space you have to travel incredible distances.

It has been an immensely enjoyable book to write. No matter how much you think you know there is always so much to learn, as I discovered on my maiden writing journey.

I felt compelled to write in a style that reflected the experience I had on my journey, so if you find it sometimes reads like I am answering my own questions, well, that's exactly how it was for me.

I hope that some who read this will go on to discover more about these subjects which I have really only just touched upon.

CHAPTER ONE

The Power of the Sun

"Every morning, the rising sun invites and inspires us to begin again."

– Debasish Mridha

The large orange/yellow sphere you see in the sky during the day is our sun.

Our sun is a star, and it happens to be a second-generation star, by which I mean there was a star in this vicinity before our star. I will come back to how we know this later.

If you wished to visit our nearest neighbouring star, Proxima Centauri, you would need to travel around 25 trillion miles or just shy of 100,000 years. You would need to be travelling at around 24,000 miles per hour though, which is, by the way, the fastest man has travelled at this time of writing.

Our sun is one of billions of stars in the Milky Way.

The Milky Way is a galaxy, and our galaxy is a spiral galaxy, which is approximately 100,000 light years across.

We are situated in one of its four major spiral arms about 25,000 light years from the centre.

A light year is how long it would take light to travel these distances. It doesn't sound like a lot when you read it, but we will get back to these distances shortly and explain just how far this actually is.

There are estimated to be billions of galaxies in the known or observable universe. If you wanted to take a trip to our closest galaxy, which by the way would be impossible with today's technology, Canis Major Dwarf galaxy, be prepared for a journey of 236,000,000,000,000,000km or 749 million years if that's easier to grasp.

A little jaunt to the nearest spiral galaxy, the Andromeda galaxy, well this would actually be an epic voyage of over 10 billion years. About twice as long as our solar system has been around.

The Milky Way: courtesy of NASA free images

Chapter One - The Power of the Sun

Now, when you look up to the night sky, you can probably see lots of stars, about three to four thousand or maybe a few more, depending on where you live of course.

In less polluted areas, you can maybe see up to five thousand.

If you can find some binoculars, you could see maybe a hundred thousand. Grab a small telescope, though, and that number goes up into the millions.

It is estimated, there are between a hundred billion to four hundred billion stars in the Milky Way alone.

All the stars you can see, with or without a telescope, are in the Milky Way. You may be able to see other galaxies but not their stars; they are just too far away.

Let's just go back to the sentences above and get this into perspective.

The Milky Way contains between a hundred billion to four hundred billion stars! These are staggeringly large numbers.

A billion is a lot! It's so much more than a million, it's crazy. For example, one million seconds equals 11 and a half days, whereas one billion seconds equals nearly 32 years. As I said, a billion is a lot.

Incidentally, recently I read in a newspaper that the billionaire Jeff Bezos, the Amazon founder, could become a trillionaire in the near future! A trillion seconds, well that would be approximately 32,000 years!

Carl Sagan, the famous astronomer, once stated there are more stars in the universe than there are grains of sand on all the beaches in the world. This wasn't just some throwaway comment, as it's most probably very true.

So then, what makes our sun so special? Well, nothing really, it's not even a big star by comparison. It's only special to us, because we owe our very existence to it. There are stars out there much larger than our own sun, some much, much larger.

Our sun is situated at the centre of our solar system, 93 million miles from earth. It is a sphere of white, molten plasma. It may appear yellow or orange, due to the colour distortion of our atmosphere, but the sun is white.

This celestial giant is approximately 15 million degrees Celsius at its core, although rather cooler at the surface, just a mere 5.6 million degrees.

Incredibly massive, it has over one million times more volume than the earth (basically you could fit over a million earth's inside the sun). A million! It may be a smaller than average-sized star compared to others, but you are probably getting an understanding of just how big our sun is relative to our little world.

To begin to understand a little of this life-giving star, we have to go back about 4.6 billion years.

So, 4.6 billion years ago, the solar system was just a load of gas and dust floating about in what's called a solar nebula. You've probably seen some of these in books or magazines. They have cool names like the Butterfly Nebula, Crab Nebula or, my favourite, the Lemon Slice Nebula which really does resemble a slice of lemon.

The Lemon Slice Nebula: Hubble images

All this gas and dust was pulled together by gravity. The majority, 99.9 per cent, of all the dust and gas from this cloud formed the sun. In a later chapter, I will attempt to explain how this happens.

The rest of it was ejected out into space, when the sun started fusing hydrogen into helium for the first time. Oh, by the way, that 0.1 per cent leftover dust and gas, well, that's what all the planets and meteors, asteroids and comets are made up of. Oh, and you and me and your neighbour's dog, in fact, absolutely everything.

So, the sun was born, and some 4.6 billion years later, we named it.

The Greeks named it Helios, for the Romans it was Sol, and we called it the sun.

Our star not only gives us warmth, which stops our planet from freezing over, it also gives us light.

Light is a very strange thing when you think about it.

Light is made up of particles called photons which have no mass. Having no mass means it can travel at the fastest speed possible, the speed of light, which is 186,000 miles per second. If light had any mass at all it couldn't get to this speed, but we will come back to that later in another chapter as well.

It's safe to say, though, that light is very fast. In fact, it only takes about eight minutes to reach the earth from the sun (93 million miles away) or about as long as it takes me, bleary-eyed, to make my coffee and butter my toast in the morning.

Now, it's an odd thing, but you can only see light when it reflects on something; take the moon, for example. You don't see sunlight streaming through space towards the moon, all you see is the brightness of the moon when it reflects the sunlight.

It's the same for light anywhere. If you have a torch, the light will only be seen when it hits a surface. Sometimes you might see a torch beam in the air, but this is just light reflecting off moisture or dust.

So, what actually is light then? Well, there has been a debate raging for a few hundred years about whether light is a wave or a particle. Some think it's both and have called this "wave-particle duality". It's actually rather complicated, and I will leave that argument at that for now.

Simply, though, light is actually caused by the heating of atoms, like in a fire. Atoms gain a bit of energy when heated, and become unstable, and then release the energy as a little photon of light. Just worth noting that this is kind of a rule of physics, energy cannot just be brought into existence and neither can it be wiped out. It can only be transformed, i.e. in this case, from heat to light!

Light may be without mass, but it can still be affected by other things. Take water for example, light passes through water slightly slower than through air, we see this as refraction. This is why your legs look detached from your body in a swimming pool!

As you know, we don't just see white light we see colours. We observe colours by the colours that are not absorbed by an object. Each colour, within the visible light spectrum, has its own narrow band of wavelength and frequency.

An apple may seem red to us, because it absorbs all colours except red and reflects the red wavelengths back to our eye. A lime, for example, doesn't absorb green. An object that absorbs all colours we will see as black. An object that doesn't absorb any light we will see as white.

Now, if you have ever held a glass prism to sunlight, you can see that it consists of different colours. With a clever invention called the spectrograph, we can see that there are lines between the colours, and these lines are the identity of the elements in the light. It's kind of like the fingerprint of the element. That's how we know what stars and planets are made of, even if they are billions or trillions of miles away. Literally, a beam of light can tell us the make-up of a star or what elements are present. You just have to know how to look for it.

Our glorious sun has an incredible surface mass area of six billion, billion square metres, which produces enough energy, in a single second, to provide all of mankind's power requirements for many thousands of years.

Just imagine if we could work out how to harness some of that power. Maybe some huge solar panels in space? Well, actually plans are being made already by China to do just that. Envisage virtually limitless and sustainable energy. Finally, we would have a chance to say farewell to those damaging fossil fuels.

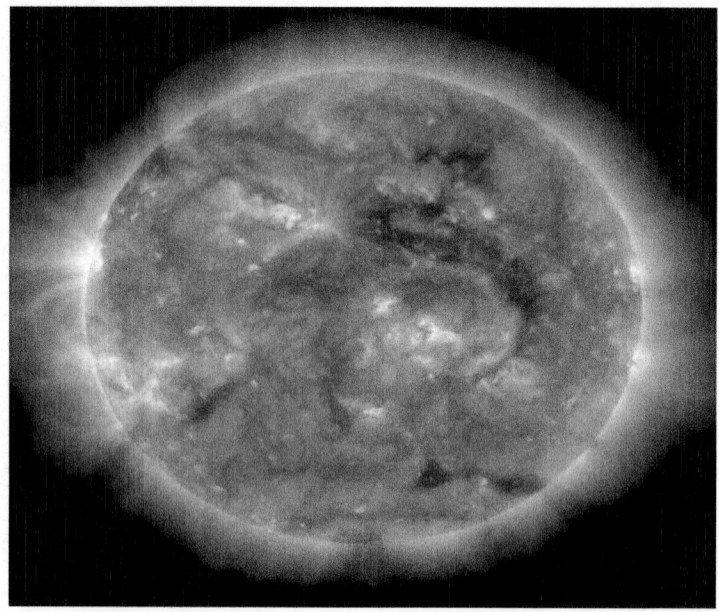

The sun: NASA free images

Chapter One - The Power of the Sun

For a long time, our star was a mystery to science. How did our sun keep shining? How could it keep producing all this energy without exhausting its fuel?

In the 19th century, German physicist, Julius von Mayer, estimated a burning lump of coal as large as the sun would only shine for about 10,000 years before using up all its burnable energy.

It was Einstein's theory, $E = mc^2$ (energy = mass times the speed of light squared), that began to unravel the mystery.

Our sun converts hydrogen atoms into helium atoms, and this is called hydrogen burning. Very basically: extremes of gravity force hydrogen atoms to squeeze together violently at incredible speeds to form helium atoms. Atoms are electrically charged and so would normally repel each other.

Until the 18th century, people still thought the age of the earth was around 6,000 years, that's what the Bible told us. Thanks in the main to geology, though, we began to understand a little of the real age of the sun and the solar system. We could see geological evidence that the earth itself was billions of years old.

But with this new understanding, the sun and its ability to keep burning became even more perplexing. The sun gave out enormous amounts of energy; despite its enormous size, it really should have run out of fuel.

It wasn't until the 20th century had given us an understanding of the radioactive decay of elements that this knowledge led us to a more accurate age of the sun; approximately five billion years.

How did they work this out?

Well, basically, some large atoms are unstable, and to become more stable they emit radiation, shedding atomic energy. This subsequent loss of radioactivity can be measured. This is normally due to the nucleus not being able to hold on to the positively charged protons, and thus the nucleus breaks apart which we measure as nuclear decay. This decay is constant over periods of time and scientists measure it as half life, basically the amount of time it takes for the measured material to be reduced by half.

This method has been used on meteorites, which is basically debris left over from the origin of the solar system. So as a result, we could accurately date the age of the sun.

Anyway! Luckily for us, hydrogen burning can last for billions of years. The sun at present is around 70 per cent hydrogen and 28 per cent helium by mass, which probably makes it middle-aged.

It's quite simply astonishing to know that our sun converts 600 million tons of hydrogen into 596 million tons of helium every second. What about the missing four million tons of matter? Well, that gets converted into energy and ejected out into space. Some of it, a very small amount, finds its way through earth's atmosphere and brightens up our day.

Unfortunately, this will not last forever, and in a few billion years or so our life-giving star will begin to exhaust its hydrogen atoms. It will then become a massive red giant (swelling out for millions of miles) engulfing Mercury and Venus. Earth may possibly escape being consumed by the sun, but our world we know today will be gone.

The oceans will have long boiled away, and the surface will be scorched and baked dry, totally devoid of any life for many millions, probably billions, of years.

Eventually, the sun will cool and become a white dwarf. It will continue to cool, for a few more billion years or so, until eventually its light will extinguish completely. With no more fuel to burn, it will just be a dark sphere of the remaining matter, not much bigger than our planet; a silent shadow orbiting the centre of the Milky Way, not quite forever.

This is not the fate of all suns, though. Stars much larger than ours, 20 times more massive, will eventually collapse in on themselves, due to gravity, and become black holes so dense that not even light can escape from the warped space around them, or alternatively, become dense neutron stars where a teaspoon of matter could weigh millions of tons.

In 1860, some German astronomers were surveying stars at the Bonn Observatory. During this process, they happened to catalogue a star which they rather dully referenced as BD-12°5055.

One hundred and fifty years later, astronomers decided to have a look at this catalogued star using the rather aptly named VLT (Very Large Telescope) which can be found in the Atacama Desert in Chile.

BD-12°5055 could be found in the not very aptly named (since this is exactly where they were looking) "zone of avoidance".

What they discovered was just simply mind-blowing. This was no ordinary star. This was an absolute stellar behemoth. It has since been named UY Scuti. To date, it is the biggest star we have ever found.

It is classified as a red giant or supergiant. To get an understanding of how big this star really is let's compare it to our sun.

Our sun has a radius of 432,450 miles. UY Scuti has a radius of 735,250,000 and has over 20 billion times more volume. When it comes to brightness, our sun is a mere candle compared to this beacon which is 340,000 times brighter.

Still not getting a perspective?

OK, try this.

If you had a plane similar to one on earth travelling at a similar speed, you could travel around the earth in just under two days. You could cruise around our sun in 205 days. If you were to take a trip around UY Scuti, though, you better make sure you book business class, as it would be your last journey. It would take around 1,000 years to travel around this immense star.

It's possible this giant has used all its helium and may turn supernova. Whatever its fate, it is, or was, an incredible leviathan of the Milky Way.

Don't be surprised, though, if they find even bigger stars out there someday. In fact, UY Scuti maybe the largest star but it's not the most massive. A star recorded as R136a1 is 265 times more massive than our sun and 9 million times brighter!

Chapter One - The Power of the Sun

Our sun's effect on space causes all the planets in the solar system to orbit around it. Not just the planets though, comets, asteroids, in fact everything, reaching out for billions of miles.

So, we know all the planets orbit the sun, but the sun, dragging the earth and all the other things in the solar system with it, also orbits the centre of the large cosmic sink we call the Milky Way.

At the centre of our enormous galaxy is a supermassive black hole, known as "Sagittarius A*". Here, it resides like a giant galactic plughole, billions of times bigger than our sun, and formed by the collapse of incomprehensible amounts of cosmic particles and hydrogen.

The sun and therefore the earth are travelling around 430,000 miles per hour around the Milky Way. Even at these fantastic speeds, it takes around 230 million years to complete one rotation. The last time we were at the position in the galaxy we are today, dinosaurs were roaming the earth and our nearest relative was probably a small rodent, like a mouse or a guinea pig, or something very unlike what we look like today.

The sun is truly a magnificent thing. Spare a thought when you look up at it. It is huge, it's been burning a long time and every living thing owes its life to it.

It does seem hard to comprehend these speeds and distances, though, and to think that something so far away could affect the movement of our little, blue planet we call home.

What is the culprit for these astonishing effects? Gravity.

CHAPTER TWO

Of Gravity, Mass and the Speed of Light

"Gravity explains the motions of the planets, but it cannot explain who sets the planets in motion."
— **Sir Isaac Newton**

On a winter's day on January 24th, 1684, in a coffee house in London, three gentlemen were having a debate about the planetary motions around the sun and what kept them in orbit.

All three men were members of the Royal Society, based in London, which was formed in 1660 for the advancement of science and nature.

These men included Edmond Halley, who predicted the return of the comet which was later named after him. Incidentally, Halley didn't discover the comet, it had been recorded at various times in history prior to Halley's observations, he just predicted it was the same comet we were seeing every 76 years.

The second was Robert Hooke, who discovered the cell and the laws of elasticity.

Finally, the third was the great Sir Christopher Wren: astronomer, anatomist, mathematician, physicist and, of course, architect of St Paul's Cathedral.

It was these three men who were pondering the problem, so the debate was in very good hands indeed.

However, none of them could seem to find a solution (although Hooke did claim to know the answer but refused to tell them). Wren was so intrigued, he decided on a wager to the first of the group to solve the problem.

Halley had an idea who could solve the dispute and set out to Cambridge University to find the great Isaac Newton.

Sir Isaac Newton was born prematurely on Christmas Day 1642. He was a weak and sickly child. It was touch and go if he was to survive, and many thought he wouldn't. Thankfully, for the scientific world, he was to prove them wrong.

With his father already passed away, Newton was soon to lose his mother as well when she remarried the local minister, the Reverend Barnabas Smith. Barnabas, it appeared, wanted a wife, but it was evident he wasn't keen on the quiet, delicate infant.

Newton's mother was to leave the family home and Newton behind to move in with her new husband. At three years old, he was unwillingly raised by his grandparents. This left an emotional scar on Newton, and it resulted in deep resentment throughout his life.

It was a time of great unrest. The English Civil War was in full swing, and it must have been a bloody and violent period to have grown up in.

Despite his traumatic beginnings, Newton eventually got to Cambridge University, where he excelled, and within a few years had become the leading mathematician in Europe.

Despite Newton's genius, he was also a decidedly odd character. He was a very private man and had little or no social life, preferring his own company to that of others. He was also prone to fits of extreme outrage.

He was deeply religious but was capable of some very strange actions. He once inserted a needle into his eye and poked around to see what happened when he distorted his eyeball. Another time he stared at the sun to test a theory on how his eyes reacted to light, making himself temporarily blind. Luckily, and somewhat fortunately for us, he escaped both with his eyes and vision intact.

He was obsessed with alchemy, turning base metals into gold, and certainly dedicated a lot of time to this, in vain of course.

He was frustrated with the mathematics of the time, so he decided to invent a new mathematic approach, the calculus, which looks at the rate of change, but then told no one about it for 30 years!

Halley's intuition was rewarded, though, when he turned up at Cambridge to meet with Newton.

When he asked about the problem of the planetary orbits,

Chapter Two - Of Gravity, Mass and the Speed of Light

Newton stated he had already solved the problem and told Halley the orbit was elliptical due to the inverse square law.

Astonished, Halley pressed him for the papers to prove it. Unfortunately, Newton had lost them! Halley begged him to redo the works.

Newton obliged and produced one of the most important scientific papers ever published: *Philosophiæ Naturalis Principia Mathematica* in 1687, or the *Principia* as it's more commonly known.

The scientific world had an answer not just to the planetary orbits but nearly all movements in the universe, including gravity and the laws of motion.

Newton became famous virtually overnight.

Sometimes the world produces men or women who are titans of science, and Newton was certainly one of the greatest examples of this. To this day, scientists still use Newton's laws for space exploration missions.

Prior to Halley's visit to Cambridge, Newton had been sent home from Cambridge to his family estate in Woolsthorpe Manor in 1665 due to the Great Plague.

Coincidentally, as I write, we are going through a similar experience with the Covid pandemic.

Newton, with plenty of time for contemplation about the world around him, had one of the greatest thoughts of all time, certainly until that point in history anyway.

He claimed he had an epiphany when he was sitting in the garden and watched an apple fall from a tree, although unfortunately this is probably untrue and romanticised for effect.

Everybody could see that things fell to earth and cannonballs, fired into the air, followed a certain path before returning to earth. Until that point, everyone had just accepted this without questioning why, which, when you think about it, is quite strange. Actually, to be fair, the great Aristotle did state that heavy objects fall downward towards the earth because of their nature and that they are returning to their natural place. Wrong of course, but at least he did think about it.

Newton, however, questioned why this should be the case. Why should something not fall up or to the side? What force is in place to make this happen?

What Newton deduced, in his time of social distancing, not only explained the above, it explained the movement of nearly everything, the tides, the orbits of the planets, it was pure genius, and this is what it states:

> Every particle in the universe attracts every other particle with a force which is directly proportional to the product of their masses and inversely proportional to the square of the distance between their centres.

OK, this might sound a little complicated, but it means that everything is attracted to everything, with a force depending

Chapter Two - Of Gravity, Mass and the Speed of Light

on its mass and distance. As you sit there reading this book, you are attracting, with your mass, the book towards you and the book is attracting you to it. You attract the book towards you more because your mass is greater, but the book still attracts you, though in a much smaller way, due to its mass. The further away you get the less the attraction becomes.

Of course, these are very small forces, so they go unnoticed to us in everyday life but, on a bigger scale, the effects are enormous.

Due to the mass of the sun, all the planets, comets, asteroids etc. are bound to it, permanently forced into an orbit around the solar giant, just as moons are held in orbit around the planets.

Newton had explained almost all the movements in the universe.

It was a monumental scientific breakthrough to relate an object falling from a tree to the same force that keeps the moon in orbit around the earth.

The power of a human mind to come up with this, just by thought alone, is really quite astonishing.

The apple fell down from the tree because it was attracted to the mass at the centre of the earth, and only the ground stopped it from going there.

Even planets, they did not know about, could be found by the effects their gravitational force had on other planets. They couldn't see them, but if Newton's laws predicted they were there, then sure enough, if they looked long and hard enough, they would find them.

So that's gravity then.

Well, actually, no!

There was a problem in that it couldn't explain some motions in the universe. Mercury's orbit didn't stand up to Newton's laws, which was strange because most other things did.

Newton's laws explained a great deal, but they didn't actually explain why mass had this effect.

It took 250 years for someone to work that out. And that somebody was a very special person indeed!

> *"Two things are infinite: the universe and human stupidity; and I'm not sure about the universe."* – **Albert Einstein**

On the 14th of March, 1879, Albert Einstein was born in Ulm, Germany. Einstein's family moved to Munich shortly after and this was where he was schooled as a child. There was certainly no obvious sign of the genius of Einstein in his early years. Even during his school years, he was considered a poor student. He struggled with languages, history and most other subjects.

A friendly Jewish student, though, called Max Talmey, befriended Einstein and gave him a book on geometry which seemed to spark his interest in mathematics.

Despite this, he was still to fail the entrance exam to the Federal

Institute of Technology in Zurich, though he was later admitted after obtaining a diploma in Aarau, Switzerland.

Several attempts at a teaching position were declined by his peers. Eventually, and in some desperation, he accepted a position as a patent clerk in the patent office in Bern in the summer of 1902.

This was a bit of a stroll in the park for Einstein, but it did offer him the opportunity to study and do some thinking. Oh, and yeah, change the world we live in and explain pretty much every movement in the universe.

In my opinion, he was the greatest scientist that has ever lived. If Newton was a titan, then Einstein would have achieved deity status. His incredible theories are still being proved right today.

Einstein had the ability to think outside the box, to ignore what science had come before and to think of the problems uniquely without any past theories to muddle his thinking. It is perhaps unfair to say that Einstein didn't have help to reach his theories, he did. He was unique, though, in the way he would think about these problems.

He would imagine himself travelling on a light beam or witnessing two lightning bolts simultaneously striking objects and how this would appear to observers, one observer stationary and the other moving.

Einstein published two papers, one in 1905 *The Special Theory of Relativity* and in 1915 *The General Theory of Relativity*.

In special relativity, we get the famous equation $E = mc^2$ (regarded by some as the most powerful equation ever written down). I'm not going to try and explain it in detail, that will be for another book, but I will try and glide over it as gracefully as possible:

E is energy, and m is mass and c^2 is the speed of light squared.

The speed of light, we know, is a big number, so imagine it squared! What Einstein's equation is telling us is that because it's such a big number even a very small amount of matter, or mass, can be converted into enormous amounts of energy and vice versa. This is how atomic bombs work.

So, nothing with mass could reach the speed of light, as it would create more energy the faster it went, in effect creating more mass slowing it down.

Special relativity has also shown us that time is relative to the observer and not constant at all, as Newton had thought.

Einstein theorised correctly that the speed of light would always be constant, no matter how you observed it in the vacuum of space, even if time had to slow down to make this happen. It's called time dilation and is very real. It's hard to grasp, but time really does slow down for you the faster you travel!

So special relativity is very complicated to understand initially, and here's what Einstein had to say about it:

It followed from the special theory of relativity that mass and energy are both but different manifestations of the same thing

— a somewhat unfamiliar conception for the average mind.

OK, so now let's get to the general theory of relativity, which is more complicated, but I believe easier to understand! It's his theory of gravity.

In 1915, Einstein published his general theory of relativity.

It is a masterpiece, and again, I will do my best to explain in layman's terms what it means.

$$R_{\mu\nu} - \tfrac{1}{2} R g_{\mu\nu} = \frac{8\pi G}{c^4} T_{\mu\nu}$$

Imagine, if you can for a moment, that space is not just a black nothingness but something spongy and elastic. OK, and then imagine something massive, like the earth, sitting in the middle of this black, spongy substance.

So, what happens? Of course, the elastic, spongy substance moulds around the earth and distorts the area around it.

Well, this is what happens in space. The earth (or any planet or star) sits in this elastic space, warping the space around it. Imagine a heavy, iron ball in the middle of a trampoline and you're kind of there.

This effect results in the moon being caught in the curvature of space around the earth. Without this effect (curvature), the moon would be gone or would have crashed into the earth.

The moon is caught rolling around the rim of the warped space around the earth. Like a penny if you rolled it around the top of a curved fruit bowl but without it ever falling in.

The reason the penny falls to the bottom of the bowl, after a short while, is due to friction slowing the penny down. No matter how fast you roll the penny, it will eventually slow down and make its way to the bottom of the bowl.

It is the same for the moon, except there's no friction in space to stop the moon from racing around the earth. So, although the moon is falling towards the earth, due to its speed, it keeps missing. This is how satellites work, it's the speed that stops them crashing back to earth. Astronauts on the International Space Station don't float around due to the gravitational warping effect of the earth; they drift around mostly due to the fact that they are actually falling back to earth, due to the speed of the space station though it keeps missing the earth.

Einstein realised, gravity, space, time and matter could all be connected or part of the same thing.

Complicated isn't it? Well, it gets even stranger!

Einstein had predicted, in the special theory of relativity, that time is not fixed, it's not constant, and it can change by how it is viewed. Newton, through his equations, presumed space and time were fixed.

Einstein's most profound and incredible idea was that time and space were interchangeable or part of the same thing. He called it space-time.

Chapter Two - Of Gravity, Mass and the Speed of Light

Let's go back to the spongy elastic that warps when we sit something massive upon it. Right, now imagine that interwoven in these chords of elastic sponge are the threads of time: seconds, minutes, hours etc.

If something massive was affecting (warping) space, then it must be warping time.

Brilliant. And more importantly, true. It has been proven time and time again.

Nuclear clocks on planes, which have been perfectly synced with clocks on the ground, have been shown to run faster due to the effects of this space-time warping.

A clock on top of Mount Everest will run very slightly quicker than a clock at the base of the mountain. Again, this has been proven to be the case.

This was the most astounding and brilliant insight. It has been said if Einstein hadn't had this thought, we could still be searching for these answers today.

Clocks on satellites that pinpoint your location on GPS have to be adjusted slightly because gravity is slightly weaker away from earth, thus time runs slightly quicker. Without this necessary slight alteration, you would find yourself in a field or a lake or a supermarket car park or anywhere, in fact, except where you wanted to be!

Newton's laws were an incredible insight; they were even elevated to a "law status" because they were so accurate.

35

Einstein, however, gave us an understanding of the celestial movements and the effects or, if you like, the mechanics of gravity.

> *"When forced to summarize the general theory of relativity in one sentence: Time and space and gravitation have no separate existence from matter."*
> **– Albert Einstein**

As I stated earlier, in everyday life for you and I, the effects of gravity are pretty much unnoticeable. However, when you get to the very large stars and galaxies, the effects can be colossal. Take for example a neutron star, a stellar object of very small radius, about the size of a city, but with enormous density, crammed tight with neutrons.

It was probably formed by the gravitational collapse of a massive star, after a supernova explosion, where the star wasn't large enough to become a black hole.

Gravity has pulled all this matter in so densely that just a spoonful of this superdense matter would weigh millions of tons. Hard to comprehend isn't it, but again it's true. Things are hard to comprehend when we don't deal with these extraordinary scenarios in our everyday life.

We experience gravity as actually quite weak in our day-to-day lives. A holiday souvenir fridge magnet can easily outmuscle gravity.

It is gravity that holds the universe together, though. It's weak enough to let us jump up in the air but strong enough so we

Chapter Two - Of Gravity, Mass and the Speed of Light

don't drift off into space.

If you could switch off gravity, which, by the way would be a seriously bad idea, we would soon lose our moon and our oceans, and the earth itself would probably start to break apart and drift off piece by piece into space.

In answer to one of the questions in the introduction: why are planets and stars spheres?

Again, gravity provides the answer. If you can remember, from the previous chapter, about how the solar system and the sun formed, all that dust and gas were pulled together by gravity.

Well, this is exactly what happens with planets. Over millions of years, bits of dust and tiny particles begin to fuse together and, as they get bigger, they start to spin and attract more particles and then bigger particles. As these fused dust particles get larger and larger, the mass at the centre becomes stronger and starts attracting objects to its centre from all around. It does this naturally, and the best natural way to condense everything to the centre is to create a circle or sphere.

Eventually, after an extremely long time, maybe millions of years, these form the spheres we call planets and moons.

This is how the solar systems formed.

But if we want to understand, or try to understand, the beginning of the universe, we have to think very differently and go back over 13 billion years.

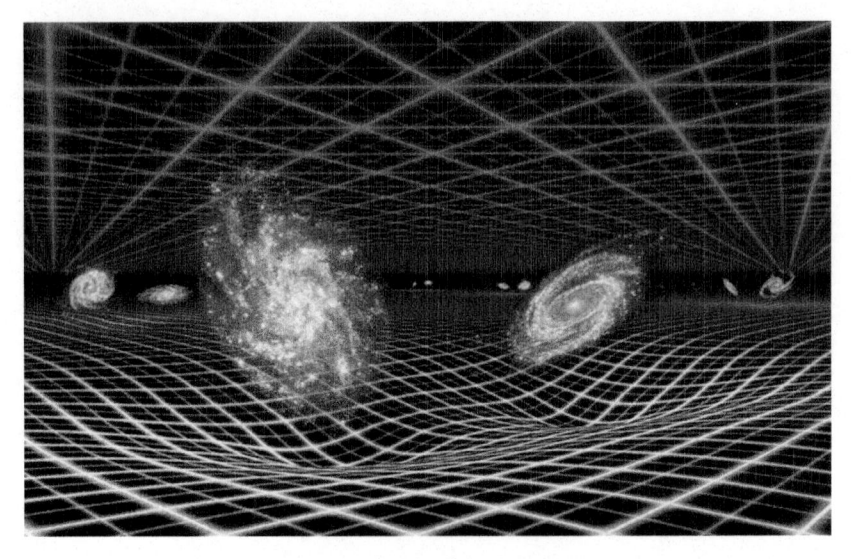

Gravity: NASA free images

CHAPTER 3

In the Beginning

"The universe is a pretty big place. If it's just us, seems like an awful waste of space."

– Carl Sagan

As Carl Sagan quoted above, the universe is a pretty big place. So vast, it's probably beyond human comprehension.

The best estimates, to date, of the size of the universe are around 93 billion light years.

Could we actually comprehend that distance if we really thought about it?

Light travels at 186,282 miles per second. And yet to get to the edge of the universe it would still take 93 billion years travelling at this speed.

That's 93 billion years!

The sun is 93 million miles away from earth, and yet light takes only a little over eight minutes to arrive on our planet.

So, it is a very, very big place. But it wasn't always like that. In fact, it was once a very small place, so small you could have maybe held it in your hand.

You are probably now thinking I've gone mad and am just making this stuff up. But if you have ever heard of the Big Bang Singularity, this is a theory of how the universe was created.

I'm going to try and explain it.

So around 13.8 billion years ago, there was no universe, no planets, no stars and also no time. In fact, nothing.

How is this possible? How can you get a universe from that? How can you get something from nothing? Well actually, you can, as long as anything you create adds up to zero.

Confused? Yep, I was too.

Thankfully, the amazing Stephen Hawking had a little analogy to help me out!

> *Imagine a man wants to build a hill on a flat piece of land. The hill will represent the universe. To make this hill he digs a hole in the ground and uses that soil to make his hill. But of course he's not just making a hill – he's also making a hole, in effect a negative version of the hill. The stuff that was in the hole has now become the hill, so it all perfectly balances out.*
> (Stephen Hawking, Brief Answers to the Big Questions)

So, when the Big Bang happened and produced fabulous amounts of positive energy, it also produced the same amount of negative energy so that the two amounts always add up to zero.

Chapter 3 - In the Beginning

Where is all the negative energy? In space, all around us.

This is, as Stephen Hawking quoted so well, *"...the ultimate free lunch".*

Getting something from nothing doesn't actually violate any of the laws of physics regarding this, apparently.

It is still hard to swallow, though, as common sense would tell us there must have been something to begin with. Well scientists are still trying to work this out, and there are lots of theories out there. Maybe the universe was born from the death of a previous universe. But then how did that universe come into being?

Maybe there are multiple universes. There had to be a beginning though, surely? Another theory is that the universe is expanding and contracting eternally and is referred to as the bouncing cosmology theory. Does this mean we continually experience all our failings and achievements forever?

If we consider the big bang singularity theory, we start with a huge explosion, or inflation, and time begins. There was no time before, no history, or past, from it to emerge from.

This seems like a lot to take in, but you have to remember that time doesn't really exist as we know it. In fact, I remember the first time I really started thinking about time: It was a day I found myself on the train, late for an appointment in London. I had been caught in a sudden down pour and I was soaked through. I had gone to my pocket and pulled out my phone only to realise the battery was dead. I had resigned myself to the fact that I was late, but now not even knowing how late. Ironically, now that I was safely undercover the sun had made an appearance and the rain that had been so intense was now barely a drizzle. It was then though I did something I hadn't done for a very long while; I looked at my watch, for the time. I had noticed it in my reflection in the train window.

It was an expensive watch. I had treasured it for years. The troubling thing though, the one actual purpose it really possessed was the one thing I didn't associate it with. I would get it out and clean it, have it repaired. Had I, at any time previously actually used it for its one solitary function? No, I don't believe I had. In fact, when I had looked at it, I could see the time it was showing had to be wrong.

All the engineering, technical invention and patient labour that had gone into making this timepiece was wasted. It had become no more to me than an expensive bracelet. My phone, my computer or even my oven were the places I would check for an instant reference, not though, the item I had purchased that was made for that sole purpose.

Time. If my watch failed to tell the correct time, then surely any clock could do the same. How do we know it hasn't? Does time only exist as we measure it? Or is it there anyway whether we like it or not? Could time be altered? Was I always destined to be late for my meeting? Was my client always going to be annoyed at my late arrival? Was there something I could have done to have avoided being late or was it there already, patiently waiting for me to play my part in an already written script?

You see, the thing is, mankind has invented time to serve a

purpose and make life easier. I manage a local football team. If I was to ask the players to meet me at some point between sunrise and midday for a match, the chances are it would be a disaster. However, if I asked them to meet me at 9 a.m., I'm pretty confident most of them would be there around this time.

This works for us humans because we think of time as linear, one straight line of seconds, minutes, hours, ticking away relentlessly. But we now know this isn't true. Time is not constant. Large areas of mass can distort time, and the faster you travel the slower time will pass for you. In fact, inside a black hole time does not exist at all due to its incredible mass. If you could stand on the edge of a black hole, which you couldn't of course, but if you could, theoretically you would be able to witness the beginning and end of the universe.

Edwin Hubble, the famous American astronomer, was the first to prove that the universe is expanding. In 1925, he detected stars that were racing away from us, and this is how we know that the universe is expanding. Note, though, it's not expanding into anything, there is nothing for it to expand into. The only space is the space it makes as it expands.

You can't get to the edge of the universe and pop your head out. In fact, even if you could travel these immense distances, it's unlikely you would find an edge to the universe, but you would possibly find yourself back where you started.

So, if we know it's expanding, we can run the process backwards, and that's how we know the universe started out as something very small or perhaps nothing at all.

Very shortly after the Big Bang, there wasn't much to begin with other than hydrogen and helium which are the most plentiful elements. Then, after around 2-300 million years, gravity brought some of this together and the first stars were born.

The birth of our solar system would have to wait a while though, eight billion years or so, in fact.

As you know, our planet has plenty of heavy elements, however, we know elements like uranium and gold, can't be formed when a star is born. So how did we get them then?

Over the course of history, people were witnessing bright lights appearing in the night sky.

The Chinese recorded one in AD 185. There were more to follow in AD 393, AD 1006 and AD 1572. These strange bright lights would last for a few months then just vanish from view.

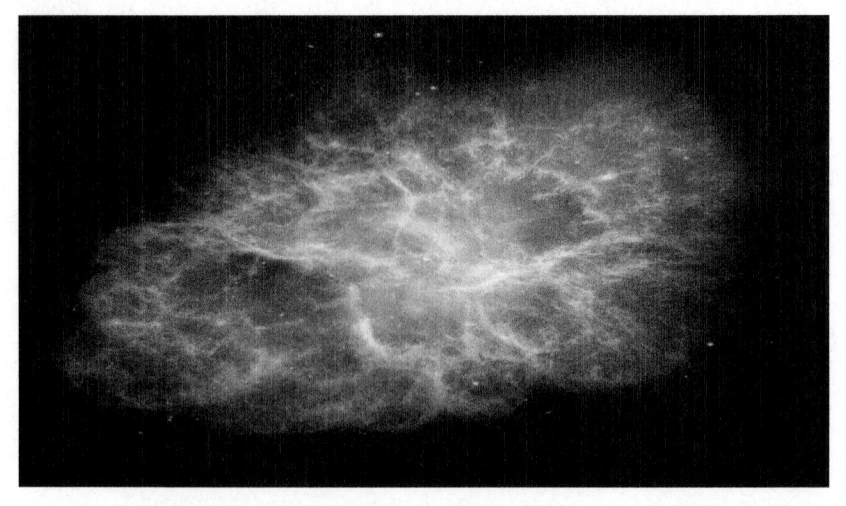

Supernova: Constellation Taurus: NASA free images

Chapter 3 - In the Beginning

It was a mystery to the astronomers at the time. We had to wait until the 1930s to get an answer.

Walter Baade and Fritz Zwicky, two astronomers, were observing the appearance of a bright light in our neighbouring Andromeda Galaxy from the Mount Wilson Observatory in California. They named it a supernova and claimed this could be the exploding or collapsing of a giant star.

Turns out they were correct and were awarded a Nobel Prize.

Now, for a supernova, as I mentioned in a previous chapter, we need a star with a lot more mass than our sun, maybe ten or more times more mass in fact. Eventually, just like our sun, one of these huge stars will exhaust all its hydrogen and helium, but as it does, heavier elements will form and build up at its centre and eventually it will implode and scatter all these heavier elements out into space.

So, all this matter that's projected out into space, yep, that's where we get most of our heavier elements from. So, we know that there must have been a supernova before our solar system formed. So, our sun is at least a second-generation star. If you have ever wondered how a rock like uranium can have energy that powers nuclear power stations, it's because locked up inside these rocks is the fizzling and crackling energy from an exploded star.

There are some theories though that suggest that some of these heavier elements, like gold and uranium, can't be put down to a supernova; they believe even a supernova can't get to the temperature required to make some of the really heavy elements. The best theories put forward for these elements are that perhaps these were formed when two black holes or neutron stars collided.

Can neutron stars or black holes collide then?

Yes, they can, and it was even predicted by Albert Einstein.

Now, as you can imagine, this would be a very violent event indeed. Probably the most violent event in the universe with the exception of the Big Bang itself.

Has this ever happened?

Yes, and what's more, we have recorded it!

Even though Einstein had predicted these colossal events and suggested that they would send out gravitational waves through the universe, he couldn't see how they could be detected or measured.

In the 1960s, though, some very clever scientists worked out that there possibly could be a way to build such a device capable of detecting these waves.

In fact, Joseph Weber from the Maryland University in North America actually predicted what sort of machine could do this.

The year 1999 saw the completion of, what can only be described as, an incredible piece of engineering: the Laser Interferometer Gravitational-Wave Observatory or, thankfully, LIGO for short.

Chapter 3 - In the Beginning

Aerial view of the LIGO detector in Livingston, LA: LIGO image

There are two detecting sites: one in Livingston, Louisiana; the other at the Hanford site in Richland, Washington.

To spot gravitational waves directly for the first time ever, scientists had to measure a distance distortion one thousand times smaller than the width of a proton.

OK, let's first get that into perspective. A proton is something very small indeed. If an atom were the size of, say, Wembley Stadium then the proton would be the size of, say, a blade of grass. So, scientists are trying to measure something one thousand times smaller than something ridiculously tiny, that has happened a long time ago and trillions of miles away.

It's pretty mad, isn't it? You just have to admire these guys who invented and work on these things.

OK, now we know they can measure these things. But just how do they measure these tiny little disturbances?

I stated earlier that massive objects colliding in space could cause a wave or rippling in the cosmos. The problem is these things happen so far away that the evidence of them, here on earth, would be so very subtle that it would be incredibly difficult to detect.

What LIGO has, is what you need to measure such things.

Basically, you need to have two beams of light travelling between two sets of mirrors down tubes that are running at right angles. A gravitational wave, or ripple, can be detected as it stretches the space in one tube and shrinks it in the other, causing the mirrors to swing slightly, which will either increase or decrease the length between the mirrors. These unbelievably tiny variations in distance are picked up by the light beams and measured.

Cool, isn't it!

This really is technology at its best. Imagine how difficult it must be to pick up a tiny ripple in space-time while avoiding detecting a ten-ton lorry driving over a speed bump a few hundred yards away.

Although LIGO didn't detect any gravitational waves between 2002 and 2010, after some advanced modifications in 2015, LIGO detected its first gravitational wave. These were thought to be the effects of a collision of two black holes 1.3 billion years ago.

Chapter 3 - In the Beginning

It's quite incredible when you think that 1.3 billion years ago two black holes collided, and that impact created great gravitational waves, like ripples in a pond, that extended for trillions of trillions of miles through space and time until they found their way to a small, blueish planet. There they entered the atmosphere of this little world and then whispered past a tiny underground, suspended mirror which was waiting for it.

It's all really rather breathtaking and magnificent.

So, just to recap, the universe is a big place. It wasn't always though, and huge stellar collisions were required to form some of the heavier elements we have here on earth and elsewhere in the cosmos.

And all that has paved the way to form our planet and our solar system, oh, and of course, you.

> "The iron that makes our blood red was made in the final moments before a star died. For all of us, then, our very lifeblood began with a spectacular death in a solar system."
>
> — **Ziya Tong**

CHAPTER 4

Of Planets and the Solar System

"You got the makings of greatness in you, but you gotta take the helm and chart your own course! Stick to it, no matter the squalls!"
— **The film** Treasure Planet
(A beautiful, animated film I watched many times with my children.)

Nasa free image

Look at the above picture, and you will probably have seen something very similar before. It shows the sun and planets of our solar system.

Chapter 4 - Of Planets and the Solar System

To scale, though, our solar system is nothing like this. There is no book big enough to show the sun and the planets to scale.

An atlas, or a globe, can be used to scale down everything to fit nicely so you can view the earth and its continents and countries easily.

For the solar system, though, it's very different.

Even if we reduced the sun to the size of a peanut, we would need a book with pages over 200 feet long to fit Neptune in.

We have mentioned before that the solar system is a very big place. Technology, however, has enabled us to visit, albeit mechanically, all the planets in the solar system including the asteroid belt which lies between Mars and Jupiter.

Well, I say all the planets... There is a theory, and a very serious one, that there is another planet out there, an absolute whopper. It's called Planet X or sometimes referred to as Planet Nine, and I'm fairly confident, if they ever find it, they will give it a more traditional name.

How, in this day and age of technology, could we miss a planet like this in our solar system? We have massive telescopes in space, and we can even find planets orbiting other stars many thousands of light years away.

So how on earth could we not find a giant planet, basically in our own cosmic back garden?

Firstly, let's answer the question about why they think there is a big planet out there. Remember back to Newton's law of gravity:

Every particle attracts every other particle in the universe with a force which is directly proportional to the product of their masses and inversely proportional to the square of the distance between their centres.

Well, this is how we found Neptune.

On March the 13th, 1781, William Herschel discovered Uranus. It was an incredible discovery, the first in fact with a telescope. Uranus is a massive ice planet, four times bigger than earth, and the only planet to be named after a Greek god (the rest are named after Roman gods).

A hundred years were to pass before two astronomers, John Crouch Adams from London and Urbain Jean Joseph Le Verrier from Paris, in 1845, both independently realised there was an anomaly with Uranus's orbit.

Something was pulling at it. They both suspected another planet. Sure enough, in June 1846, they found what they were looking for. Actually, it was the German astronomer **Johann Gottfried Galle** who was the first to observe the new planet, but it was Adams and Le Verrier who advised him where to look.

Neptune is a gas planet, and also about four times bigger than earth, but it is the smallest of the gas planets. It's an incredible 2.8 billion miles from the sun. Now you can see why it was so hard to find, even with a telescope.

Chapter 4 - Of Planets and the Solar System

It's these same calculations that predicted Neptune that have caused astronomers to think that there could well be a giant Planet X out there. There are some ice bodies in the Kuiper belt, past Neptune, where there are again anomalies in their orbits.

The theory is that a huge planet, on an extraordinary 10,000-year or more elliptical orbit, is the cause of this. The problem is, at these distances, planets are very hard to find, as they don't reflect much of the light from the sun.

Astronomers are looking though, and we can be confident that if there is a Planet X out there, one day they will find it.

So how can we see planets, that orbit other stars, that are much further away?

The answer is we can't. However, we can confidently predict they are there.

Most of the planets we have discovered orbiting other stars are found using the transit effect.

Basically, a star's brightness can be measured. Now, if there is a dimming of the star when observed, then they consider the possibility that this could be caused by a planet passing in front of it.

Working out the amount of dimming gives an indication of the size of the planet. Then after observation, we can work out its orbit by the number of times the star dims. This is a very basic description, and there are other ways to tell, like if the star wobbles very slightly, this would indicate the presence of a planet or planets.

The diagrams below, taken from NASA, give an indication of how this dimming is measured.

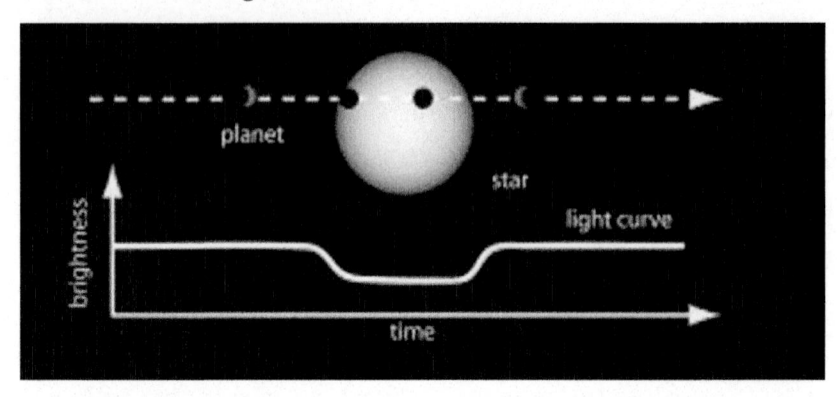

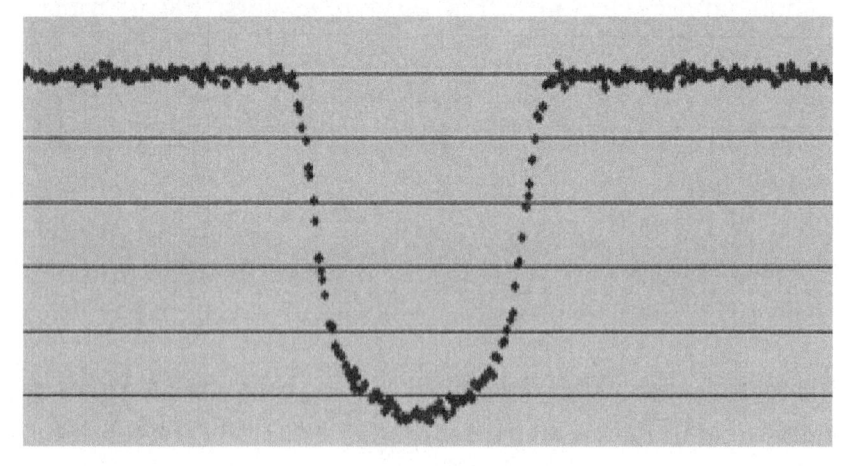

(Nasa.gov.free to all images)

Back to our solar system, though!

If, like me, you enjoy looking up at the night sky on a warm, summer night and try to identify stars, you may well have noticed that some of the brightest lights in the night sky are actually planets.

Chapter 4 - Of Planets and the Solar System

One evening, when we had friends around for late drinks in the garden, I mentioned how bright Jupiter was. They looked at me a little bemused. They all thought it was a star! In fact, they were all unaware that you could see planets and presumed all the lights in the night sky were stars.

In fact, most of our planets can be seen in the night sky without a telescope.

I have been in awe of the planets for as long as I can remember. As a child, in my local library, I searched for any books that would have pictures or descriptions of them. I was absorbed by news of space probes visiting the inner, rocky planets and then missions to the outer planets.

We have seven planetary neighbours, all of them incredible in their own unique way, and here is a just a little about all of them.

Mercury: images-assets.nasa.gov/image/

Mercury is the closest planet to the sun. It's 36 million miles from the sun, in fact, and it's the first of the four inner, rocky planets.

It's also the smallest. In fact, there are bigger moons in the solar system. So Mercury can be quite difficult to see due to its close proximity to the solar centre.

If, in the unlikely and unfortunate event, you were to ever find yourself teleported to Mercury, you would die in approximately two seconds. It can range from 450 degrees Celsius during the day to minus 180 degrees at night.

Having no, or very little, atmosphere doesn't allow Mercury to hold on to any of the heat it receives during the day, hence the extremes in temperature. You can probably get an idea of this on earth during spring, where it can be reasonably warm during the day, but the temperature soon plummets when the sun goes down.

Over 58 days on earth would need to pass to make the equivalent of a day on Mercury. This is due to its incredibly slow rotation. It may rotate slowly, but Mercury was named by the Romans after the messenger god, probably due to its swiftness in the night sky. Swift doesn't really do it justice, Mercury speeds around the sun at 100,000 miles an hour, making a year on Mercury just 88 days.

Mercury's fate, unfortunately, is to be engulfed by the sun. Only its dizzying speed stops it from being devoured. Remember the penny around the fruit bowl analogy from the chapter on gravity?

Chapter 4 - Of Planets and the Solar System

Despite being closest to the sun, it's not the hottest planet though!

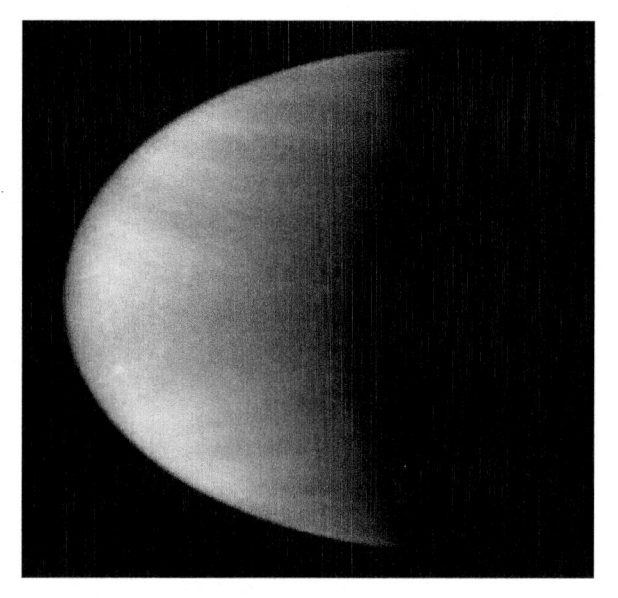

Venus: images-assets.nasa.gov/image/

Venus is the second rocky body from the sun, and daytime temperatures on Venus are around a sweltering 471 degrees Celsius. I say daytime, but actually, and rather bizarrely, a day on Venus is actually longer than a year on Venus, rotating once in 243 days compared to is 224-day orbit around our sun.

Venus orbits the sun at approximately 67 million miles and, at certain times of the year, it's the brightest light in the night sky. This possibly led to the Romans naming it after the goddess of love and beauty.

Venus can be seen during the day, and it's often referred to as the morning star or evening star, as these are the best times to see it.

Some astronomers make reference to Venus as earth's twin, due to its comparative size to earth, but this is where the similarities end. Venus may have been named after the god of love and beauty, but they did not know, as we do now, how truly hostile it is.

Venus has a very thick, toxic atmosphere, consisting mainly of carbon dioxide, trapping in heat hot enough to melt lead. One day on Venus would be nearly 117 days on earth. Like Mercury, it also has a very slow rotation.

Venus may not have always been this way though, it could have possibly resembled earth at some point; it's certainly in the right zone for water to exist as a liquid. If we ever needed a reminder of what a runaway greenhouse effect can produce, we need to look no further than Venus.

It should be sobering enough for immediate international action for our delicate planet. In fact, it's somewhat bizarre we haven't acted with more intensity sooner when we realised how toxic Venus has become! You wouldn't look to find life on Venus, not life as we would understand, that is.

In a bizarre twist, as I was writing this, scientists have found evidence of a substance called phosphine on Venus. In fact, it was originally discovered on Venus back in 1978 when NASA's Pioneer Venus probe descended into the Cytherean atmosphere.

This is a bit of a mystery, as phosphine is normally an indicator of life! Phosphine is made of a molecule of phosphorus and three hydrogen molecules and is found on earth as a by-product

Chapter 4 - Of Planets and the Solar System

of microbes. Could there be basic life forms, such as microbes, in the clouds of Venus? Well, it's too early to tell, but it has definitely drawn the attention of the scientific world. I guess there will be other explanations but, at this time of writing, it is possible that, just maybe, for the first time, we have found some small indicator of life elsewhere in the universe.

Mars: images-assets.nasa.gov/image/PIA04591/PIA04591~orig.jpg

Mars, at nearly 142 million miles from the sun, is the last of the rocky planets, and everything about it makes it unsuitable for human life.

Some billionaires, who imagine inhabiting the red planet one day, will certainly have their work cut out. We sometimes refer to it as the red planet, and that's because rusting iron minerals give Mars its red tinge.

Yes, there is oxygen on Mars to make this happen, although not a lot, and we know, thanks to some of the rovers, there was also liquid water on Mars in its distant past.

Mars is named after the Roman god of war and not a chocolate bar, in case you wondered.

It's also the second smallest planet, about half the size of earth. It can be quite difficult to see Mars, as it appears as a tiny, red dot and is best seen at its closest orbit to earth.

The Martian atmosphere is 95 per cent carbon dioxide; at these extreme levels it would be fatally toxic to a human being. Temperatures are around minus 60 degrees Celsius and, at the poles, it can go down to minus 140 degrees.

Could we colonise Mars? Maybe. It's highly unlikely to be in the near future, though.

It's worth noting that Mars is also home to the largest mountain in the solar system: Olympus Mons. It's actually a volcano and it's huge, two and a half times bigger than Earths Mount Everest.

Due to its size, gravity is about three times less on Mars, which means you could jump about 10 feet in the air; only if you wanted to, of course.

Is there any water on Mars now? Well, in 2018, they did find evidence of a possible lake under the polar ice cap! As I write this, three more missions are on their way to Mars, and we shall mention more of these later.

Chapter 4 - Of Planets and the Solar System

Asteroid belt: images-assets.nasa.gov/image/

After Mars we have the **asteroid belt**, which covers an area 140 million miles across. That's an area larger than earth's distance to the sun.

This is a large stretch of space consisting of billions of rocky formations, maybe once a planet or a moon that didn't quite make it due to the gravitational effects of Jupiter, or perhaps just remnants left over from the formation of the solar system. Most are quite small, though, but there are some very big ones as well. The largest is Ceres which is about 590 miles wide.

Science fiction films display these areas as densely populated with large rock boulders that you will need to blast your way through.

This is simply not the case for our asteroid belt. There may be billions, if not trillions, of rocky or icy objects out there in this zone, but the chances of actually hitting one, or even coming close to one, is immensely small, a chance in a billion. When sending probes to the outer planets, scientists don't even take these into consideration, as the odds of a collision are so staggeringly remote.

Every so often, though, one of these rocks gets nudged our way and smacks into earth, sometimes with devastating consequences. They can smash through earth's atmosphere at anywhere between 30 to 40,000 miles an hour.

At least three times these impacts have nearly ended all life on earth. It's not a question of if another one hits our planet, just a question of when!

Jupiter: images-assets.nasa.gov/image/hubble-capture

Jupiter is an absolute monster. Not quite having the mass to

Chapter 4 - Of Planets and the Solar System

become a star, it's by far the biggest planet in the solar system. It's so big, it's more than twice as big as all the other planets put together.

Our sun is made up primarily of hydrogen and helium, and this is also the case with Jupiter. However, to begin fusion and become a star, it would still need to be much more massive, about 80 to 100 times more massive in fact.

Jupiter is one of the brightest lights in the night sky, and it orbits the sun at over 480 million miles. It's probably had a lot to do with the solar system being the way it is, with regards to the position of the planets and their orbits. Its huge gravitational effects also influence comets and asteroids.

This Jovian Polypheme is what scientists call a gas giant. It acts like a giant shield to earth; its huge gravitational presence protects the earth from asteroids and comets but it's very plausible, in fact very possible, that Jupiter has deflected asteroids this way, with devastating consequences, before in the distant past.

In the late 1960's Jupiter captured a comet. It was huge, and began orbiting the enormous planet. It was first witnessed in 1993 by three astronomers: Eugene and Carolyn Shoemaker, and David Levy. It was named Shoemaker-Levy 9.

Astronomers soon realised that this comet was going to eventually crash into Jupiter in 1994. It was to be the first celestial collision ever to be witnessed by mankind.

Many astronomer's believed the enormous planet would just

gobble up the comet without much of show, In fact some referred to it as the big fizzle.

On the 16th of July human beings witnessed the collision. It was rather disturbing. The comet, now in fragments, smashed into the Jovian atmosphere with the impact force of 300 million atomic bombs. It was a spectacular show that lasted for days. The atmosphere was super heated and plumes were recorded to have reached nearly two thousand miles high. A big fizzle it most certainly wasn't.

Changes to the atmosphere have been detected right up to 2013 as a direct result from the impact. Suddenly Earth didn't feel as safe a home as it had before. Hollywood jumped on the chance to make it seem even more desperate with the release of world ending collision films. It wasn't long before Nasa was tasked with looking and identifying any objects that might one day come this way.

So how big is Jupiter? Earth's diameter is 12,742 kilometres, Jupiter's is over 142,000 kilometres and 1,300 earths could be squeezed inside this enormous ball of gas.

Aptly named after the king of the Roman gods, the mighty Jupiter has 79 celestial companions or moons, and we will get to some of them in another chapter.

At the current rate of our best space travel for a human, it could take you up to six to eight years to get to Jupiter, although this could be reduced by half if the planets are in alignment. New technology for unmanned probes, though, are now able to get to Jupiter much quicker.

Chapter 4 - Of Planets and the Solar System

On the 5th of August 2011, NASA launched its spacecraft Juno. On July the 4th 2016, it arrived at Jupiter.

The 66-foot-wide probe was sent to study Jupiter's core and magnetic field, among other things. It's still in orbit around Jupiter and will also go on to study the Galilean moons we will be discussing in a later chapter.

The mightiest hurricane on earth would be a breeze on Jupiter. Jupiter's red spot, which is about the size of three earths, has winds of nearly 400 miles an hour.

Due to the high carbon in the atmosphere and dense pressure, it is very possible that it rains diamonds on Jupiter. Imagine that!

Saturn: images-assets.nasa.gov/image/PIA03152/PIA03152~orig.jpg

Saturn. If you had to pick the most beautiful planet in our solar system, I bet most of you would pick Saturn. It's kind of the jewel in the solar system. It's the sixth planet from the sun and the second largest planet in our solar system. It's another huge gas giant and is also made up primarily of hydrogen and helium. Jupiter has 79 moons, however, Saturn caps that with an astonishing 82 moons!

What makes Saturn so beautiful? The fact that it is decorated with amazing rocky and icy, flat rings.

I had always wondered why Saturn had such beautiful rings. These rings are made nearly entirely of ice, and they extend over 400,000 kilometres out into space, but are only about 10 metres thick. The best theory for this is that a moon got too close to Saturn and became unstable in its orbit, as it got closer Saturn's gravity ripped it to shreds.

Sadly though, these rings, however they were caused, are likely to be short-lived. In a hundred million years or so, they will probably be gone, as Saturn's gravitational effect overcomes the velocity of the debris and claims them. The NASA image below shows the moon Pan that has cleared a gap in the rings.

images-assets.nasa.gov/image/PIA20490/PIA20490~orig.jpg

Chapter 4 - Of Planets and the Solar System

We should perhaps count ourselves lucky that we have lived in a time where these rings have been visible to us.

Saturn is named after the Roman god of agriculture, and they also named a day of the week after it as well, Saturn day, or as it's better known now, Saturday.

So how big is Saturn? Well, it's not as big as Jupiter, but it's still eight or nine times wider than earth with a radius of 36,000 miles!

We are starting to get a long way from the sun now, nearly 900 million miles. It takes light eight minutes to reach earth travelling at the speed of light but 80 minutes to get to Saturn.

Uranus as seen by Voyager 2: images-assets.nasa.gov/image

Uranus is an ice giant. We are a very, very long way from the warmth of the sun now, a staggering 1.8 billion miles away in fact.

Relatively Spacious

The seventh planet from the sun, Uranus is also the third biggest.

Strangely, Uranus spins on its side, which is unique to the planets, so it kind of rolls around the sun.

It was, as we found out earlier, the first planet found with a telescope. Uranus was discovered in 1781 by British astronomer William Herschel (Herschel was actually born in Hanover, Germany in 1738), although he originally thought it was either a comet or a star.

Herschel, once he realised what he had discovered, tried, unsuccessfully and thankfully, to name his discovery Georgium Sidus after King George III.

Uranus is a lovely bluey-green colour which is probably caused by large quantities of methane gas in its atmosphere.

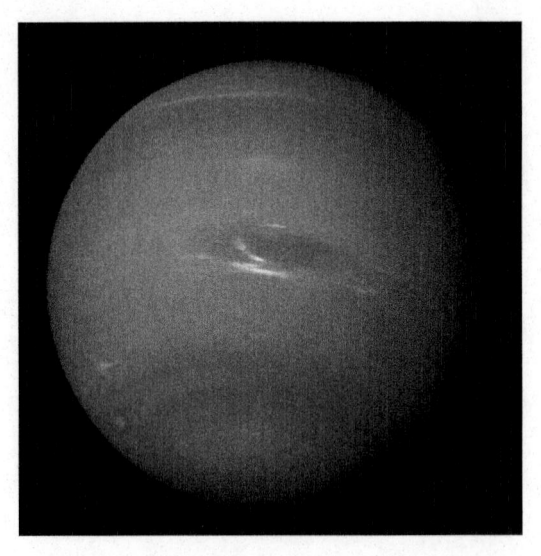

Neptune: images-assets.nasa.gov/image

Chapter 4 - Of Planets and the Solar System

Neptune is the most distant planet in the solar system, named after the Roman god of the sea.

We are now 2.8 billion miles from the sun. That's 2.8 billion miles! It would take you about 100,000 years to walk there, if you wanted to that is. Actually, that's nonsense, you couldn't walk there, but you could travel in a rocket. The fastest one we have that has successfully carried a human being would take you about 15 years. Even travelling around 24,000 miles an hour. You are probably getting an understanding of just how far away Neptune actually is. It takes Neptune just under 165 years to complete one orbit of the Sun. So only one year has passed on this distant planet since we have discovered it.

Neptune is a very cold and dark planet. Sunlight takes nearly four hours to reach it. Temperatures here can get down to minus 220 degrees Celsius. It's a very inhospitable place with great storms continually raging on this remote world. These magnificent winds can reach more than 2,000 miles an hour.

Neptune also has rings, but they are mainly made of dust and rock and so do not look as impressive as the rings of Saturn. We have only managed to visit Neptune once and that was back in 1989 courtesy of NASA's Voyager 2 space probe.

This desolate, gaseous and icy world isn't the end of the solar system though. If we regard the solar system as where the effect of the sun's gravity is still in place, then there is still a very long way to go.

After Neptune we will find the Kuiper belt, where we will find millions of icy bodies including the former planet Pluto.

Beyond the Kuiper belt, well billions of miles beyond, is the Oort cloud. Scientists haven't seen this place, and it's totally hypothetical, but this is where they think some of the long-orbit comets come from.

So, briefly, that's the solar system and its planets. There is so much more to find out about them though.

Our neighbouring planets have been with us for billions of years and will be for billions more years.

I find it incredibly exciting to think that, one day, mankind may actually visit one of these planets. Despite the many formidable hurdles, I'm convinced we will.

I know we will try.

CHAPTER 5

Amazing Earth

"Earth provides enough to satisfy every man's needs but not every man's greed."

– **Mahatma Gandhi**

You probably noticed that I missed out earth in the chapter regarding the planets. This was because I thought our home planet deserved a chapter of its own. Obviously, a whole volume of books could be written about the earth, so I'm only going to just lightly touch on a few relevant, interesting facts that you may, or may not, know about our planet.

Earth is the third planet from the sun, 93 million miles from the sun in fact. We orbit our star at 67,000 miles per hour.

OK, let's think about that for a moment. At 67,000 miles per hour, that's how fast we are travelling through the void of space.

Over 30 times faster than a bullet fired from a gun. Yet we don't feel the effects of this.

Well, it's the same principle as flying, or travelling in a train or a car. When you are seated on a plane, you don't feel as though you are travelling at hundreds of miles an hour because you are on the plane, so you are relative to its speed. It's the same as being on earth; we don't feel the effects because we are travelling at this speed as well.

If a visitor from another part of space was looking for a good place to live in our solar system, earth would be a good place to start looking. Presuming, of course, they were a carbon-based life form that breathes oxygen and relies on liquid water for its survival.

Earth is in a part of the solar system that we call the Goldilocks zone. Not too close to its parent star that its water would boil away, and not too far away where liquid water would freeze. It is in fact just right! Just like the little bear's porridge.

Astronomers, looking for life on other planets in the galaxy, look for planets in the Goldilocks zone, as they are the most likely to be suitable for life. Life, as we know it, that is.

As you probably know, the earth's surface is roughly 70 per cent water, and our best guess to date is that the water was deposited here by comets in the late heavy bombardment (a time about four billion years ago when the solar system was bombarded by asteroids and comets left over from the forming of the solar system). About 96 per cent of all this water is in the oceans, with over 50 per cent in the Pacific Ocean alone. Only about three per cent of all the water on earth is freshwater (most of which, about 70 per cent, is caught up in ice caps and glaciers) and only a small per cent, a little over one per cent, can be used as drinking water. Water is all around us though, in the air, in the ground, in you and your pets, in fact, in everything living.

Chapter 5 - Amazing Earth

In an extremely fortunate but quite bizarre way, water, as it freezes and turns into a solid, also becomes lighter! Ice floats, as you well know.

This is quite extraordinary, though, and isn't what you would expect. Normally, as something becomes a solid, it contracts and becomes heavier. Well, as I said, this is extremely fortunate because probably without this effect, I wouldn't be here to write this book and you wouldn't be here to read it.

We know, through geological records, earth has completely frozen over a few times in its history, it's called Snowball Earth.

If ice had been heavier, it would have sunk to the bottom of the seas and lakes and it would have frozen upwards making it completely solid. The shining ice would have reflected back the sun's rays, and it's possible, in fact highly probable, that earth would have remained a frozen, dead world!

Why does ice have this weird effect? Well, this happens because when the molecules in water (H_2O) get below four degrees Celsius, they start to slow down and crystallise which reduces its density, which actually happens to be very convenient for all living things on earth. So, ice is pretty cool!

As you probably learnt from school, the centre of our planet is home to a giant, solid, iron core. Surrounding this core is an ocean of liquid metal. This again is extremely fortunate for us, because this movement of the liquid around the core is what gives the earth its magnetic field, a bit like a bicycle dynamo effect.

The magnetic field protects earth from the sun's harmful radiation and cosmic solar rays. You can see the magnetic field in action at both the north and south polar regions, where the solar radiation hits the magnetic field and creates amazing colourful auroras.

The magnetic field, as powerful an ally to the earth as it is, cannot protect us from all the sun's destructive rays though. Luckily, we have another protective shield, our atmosphere.

Earth's atmosphere is enormous, it reaches out so far it even affects the International Space Station's route.

So why does earth have an atmosphere? It's a good question and it's very complicated.

When earth formed, about 4.5 billion years ago, the molten planet barely had an atmosphere at all. But as our planet eventually cooled, its atmosphere formed. Popular theories suggest that this early atmosphere was delivered from volcanoes in the form of carbon dioxide, methane and nitrogen. Hardly the gases we would associate with being good for life!

Somewhere though, maybe in the oceans, life did get a foothold. Very basic life of course, but at some point, perhaps a couple of billions of years later, this basic, early life started converting the sun's energy into oxygen.

This obviously took a very long time, but eventually, it completely transformed our atmosphere.

The gases trapped in our atmosphere act like a big solar

panel stretched all around the earth. Here, the sun's energy is collected, absorbed then released in a greenhouse effect back to earth.

The earth would be a very cold place, perhaps frozen, without this protective layer of gases.

Unfortunately, as the human race deposits more carbon dioxide into the atmosphere, the more this greenhouse effect runs out of control, resulting in the warming of the climate, which will eventually have catastrophic effects. Take a look at Venus!

It is gravity, by the way, that holds on to this protective shield.

So, very basically, we absolutely need our atmosphere, as it protects us from the harmful rays of the sun, and we need to make sure we take very good care of it.

The earth's orbit around the sun is not circular; it's actually elliptical or kind of oval.

Now, for a long time, I presumed that when the earth is nearest the sun in this orbit, we would experience summer and when the earth is furthest away in this orbit, we would experience winter.

Turns out I was wrong. It's actually the earth's tilt that causes the seasons. The tilt was probably caused by a mighty collision that I will elaborate on more in the next chapter. Anyway, scientists think it was this collision that knocked the earth off its axis.

It's another one of those very fortunate events that most

probably helped us to become us.

Imagine a straight line, or stick, going right through the middle of our earth, sticking out from the North Pole to the South Pole. The earth spins around this pole, making one complete rotation each day, or every 24 hours, if you like. This is why we have day and night.

When we are facing the sun, it is daytime and when we are not it's night-time.

OK, that might be really obvious. However, this tilted axis is why earth has seasons, it's due to the fact that the earth's axis doesn't stand up straight.

So basically, depending on which pole, north or south, tilts towards the sun during its 360-day orbit, will depend on which hemisphere experiences summer and which experiences winter. Obviously, after six months, the effect is reversed. The time in between will be spring and autumn. Places at the equator stay hot all year round, because the tilt doesn't really affect the centre of the tilted sphere.

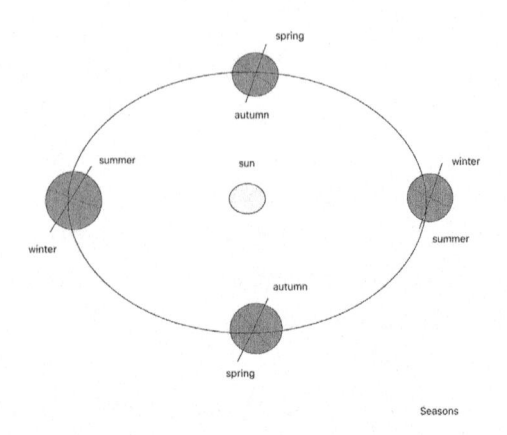

Seasons

Chapter 5 - Amazing Earth

We have mentioned the atmosphere, the magnetic field, the seasons and even the earth's core, but what about the very ground beneath our feet?

I'm sure, at some point in your life, you will have seen a globe or an atlas showing all the continents and countries spread out around the globe. This wasn't always the case though. A few hundred million years ago, there was just one supercontinent called Pangaea.

The idea that the continents moved, or slid, around the surface of our planet seemed ridiculous to most scientists when it was first proposed.

In 1915, a guy called Alfred Wegener, a German meteorologist and geophysicist, published a paper explaining that islands or continental land masses were moving or sliding across the earth's crust. Sometimes, they would smash into each other. He named it continental drift. Wegener was certain, like other scientists before him, that the coastlines of South America and West Africa had, at one time, been joined together. When you look at the similarities, it is quite easy to see how he and others had come to this conclusion.

Actually, he was right, and it was a really good theory. Where Wegener fell down, though, is that he couldn't explain why this happened. What force caused this to occur?

He proposed that it was maybe the earth's rotation that caused this. Actually, he was wrong.

Today, we call it "plate tectonics", and we know continents

sit on enormous rocks, or slabs, that are always on the move. Simply, this is due to molten rock in the mantle being superheated by the earth's core. We call it convection movement, where heated matter rises and then lowers again when it cools down.

This is still happening today, and tectonics have caused massive mountain ranges like the Himalayas, as these enormous Indian and Eurasian slabs literally smashed together forcing millions of tons of rock upwards to create this 1,800-mile-long mountain range.

The continents of Africa and South America are still moving away from each other, as are North America and Europe. Only by a few centimetres each year though.

So, whilst we humans go about our everyday lives, blissfully unaware of the movements of massive slabs of rock beneath our feet, we are also unaware of the movements and goings-on in the solar system and the fact that huge amounts of stellar dust descend down upon our world. It goes unnoticed to us, which is rather strange, because it's estimated to be in the region of 100 tons every day!

This cosmic debris is made up of tiny particles probably from comets or icy bodies that have begun to lose mass as they get closer to the sun.

We may be blissfully unaware of what goes on under the ground and above the atmosphere, but things happen on our surface which defy belief, as you will see.

Chapter 5 - Amazing Earth

If you ever find yourself in the unlikely event of being a passenger in a space vessel far from earth, the one thing I am sure you will not do is smash up or break the machine that gives you oxygen; it would be a ludicrously stupid thing to do.

And yet, here on earth, the very things that give us oxygen and remove poisonous gases from our atmosphere are being removed by man at an alarming rate.

This book isn't intended to be about climate change, but it's worth noting that in 2019 around seven billion trees were felled; mostly for areas of cattle farming.

Seven billion trees!

It doesn't take a lot of intellect to work out that this cannot be sustainable. We do plant trees, but no where near enough to replenish what's been removed.

The World Wildlife Fund estimates that the Amazon rainforest has lost 17 per cent of its trees in the last 50 years.

These are delicate ecosystems that will have serious global repercussions if not addressed very quickly.

Human beings have harnessed many of the incredible forces on earth, unfortunately, and almost sneakily, we have also gifted ourselves the ability to destroy them.

Relatively Spacious

The English dictionary describes the word parasite:

> *A person who receives support, advantage, or the like, from another or others without giving any useful or proper return, as one who lives on the hospitality of others.*
>
> (Dictionary.com)

Kind of disturbing, don't you think?

Earth is our home, and it's most likely to be the only one we ever have. Although it does seem very fortunate that all the above have made life possible on this little planet, it shouldn't be taken for granted. In a later chapter, you will see just how precious life is and also how fragile a hold we have on it.

Earth: images-assets.nasa.gov/image

CHAPTER 6

A Bit about Moons

"Don't tell me the sky is the limit when there are footprints on the moon."

– **Paul Brandt**

Leaving the earth's atmosphere is quite a challenge. To overcome the effects of earth's gravity, you need to be travelling at a certain speed.

This speed is called escape velocity, and this is worked out by an equation. Every planet and star will have one. The earth's escape velocity is a staggering seven miles per second. That's around 25,000 miles per hour.

We can use the same equation $v_e = \sqrt{(2GM/r)}$ for any size of object. Basically, it means escape velocity is governed by the radius of the planet, the mass of the planet and Newton's constant gravitational laws.

So, the bigger and more massive the planet or star, the faster you will need to travel to leave it behind. Imagine trying to run away from something whilst you are attached to it with a piece of elastic and you will kind of see what I mean.

Relatively Spacious

OK, let's try some other planets. How about Jupiter? It's much more massive than earth and with a much larger radius. So, using the equation with Jupiter's mass etc. we know we would need to be travelling at just over 37 miles per second.

OK, let's try the sun. How fast we would need to travel to escape the surface of the sun?

You would need to be travelling at 400 miles per second or 1,440,000 miles per hour. That's like travelling from the northern most point of Scotland to the southernmost tip of England in a little over a second.

This same principle works for black holes, because their mass and radius are so large even light, travelling at over 186,000 miles per second, still can't break the escape velocity, hence no light can escape and that's why they are called black holes.

So, back in the 1960s, when NASA boldly claimed they were going to send a man to the moon, this was a very courageous statement indeed. To reach the speeds required, NASA built the *Saturn V* rocket, and it stood over 360 feet tall. It is, to this day, the biggest rocket that's ever been made.

At 9.32 a.m. on the 16th of July 1969, *Saturn V* launched from the Kennedy Space Center. Over seven million pounds of thrust jettisoned Neil Armstrong, Michael Collins and Buzz Aldrin into space.

Four days later, they landed on the moon

Chapter 6 - A Bit about Moons

The first human beings to ever set foot anywhere other than earth.

The rocket parts that returned through the earth's atmosphere, after sending these pioneers to the moon, crashed into the ocean at around 5,000 miles per hour. Incredibly, and rather delightfully, these have been located despite the fact NASA had resigned them to the ocean floor forever. The Amazon founder Jeff Bezos located them and raised them from the seabed.

They are now being worked on for potential future missions or, possibly, they may just end up being displayed in a NASA museum. Either way, I think it's rather cool and befitting of this immense rocket, rather than rusting away, forgotten and discarded, at the bottom of the ocean.

Saturn V needed nearly two million litres of fuel to achieve the thrust required to get into space. It used 20 tons of fuel per second. Imagine sitting on a small capsule attached to the top of it!

Well, we know it was a success, but we also know that when there is a problem it will almost certainly end in death for the poor astronauts.

Our moon is 239,000 miles away. Probably one of the first things you will notice is the number of craters on the moon; it looks like it's been used in some crazy asteroid pinball game.

Scarily, the earth would look the same had we not had constant geological movements and vegetation.

So how did the moon come to be here?

Well, it's a good question. The most likely theory now is that a Mars-sized planet, called Theia, caught the earth with a glancing blow in its early formation. This impact incinerated Theia and turned it into rubble.

This gigantic pile of space rubble, caught in the earth's orbit, was eventually pulled together by gravity to form a moon.

It's probably a good job for us it did, as it's likely this impact caused the earth's tilt which gives us our seasons!

We know we owe our tidal forces to the moon, but we can also thank the moon for the length of the day, as its gravitational effect on the earth slows the spin of our planet down. Without the moon, we could have days lasting as little as nine or ten hours.

Imagine having a night with no moon or no moonlight? Earth at night would have been an incredibly dark place before we had mastered fire.

The moon rotates on its axis at the same rate as the moon orbits the earth. This phenomenon is called synchronous rotation or tidal locking. This is why we only ever see one side of the moon. The dark side isn't always dark though, it gets plenty of sunshine, it's just because of its tidal locking that we only ever get to see the side that's facing the earth.

We see the moon clearly, because when the sunlight strikes the moon it reflects its light.

Chapter 6 - A Bit about Moons

Over the period of a month, the moon will orbit the earth and we will see which parts of the moon are lit by the sun from our perspective.

Half of the moon is always lit by the sun, but we cannot always see it because that lit half may not always be facing the earth. We call these observations the phases of the moon.

By a galactic coincidence, the moon is at the exact right distance to cover the circumference of the sun.

The moon is 400 times smaller than the sun, but the sun is 400 times further away.

This cosmic bit of fortune allows us to see a solar eclipse. The eclipse happens when the moon lines up with the sun, blocking its light.

I am lucky enough to have witnessed one back in the nineties, and it was a really weird experience. One moment it's the middle of the day, the next it goes dark and all the birds stop singing. Within a few minutes, everything is light and back to normal again. It must have seemed very strange to humans thousands of years ago.

A lunar eclipse is when the earth's shadow directly passes in front of the moon. This is where we get the beautiful, red, fiery colours of the moon. In fact, it's often referred to as a blood moon.

Could life have evolved on earth without the moon? Possibly, though probably not as we know it today.

Relatively Spacious

With the advantage of space travel, we can now be pretty confident regarding most of the moons in the solar system.

Mars has two moons, Phobos and Deimos. Both are quite small, less than 10 miles wide. These are likely to be captured asteroids in Mars's distant past. Phobos spins quickly around Mars, and it's likely that it will eventually crash into Mars in a few million years.

Jupiter has 79 moons and is also home to the biggest in the solar system. They are all fabulously named after beings seduced by the mighty king of the gods, Zeus.

I am not going to mention them all here, but I do feel compelled to mention a few of my favourites. Before we do though, we need to know a little of the great astronomer who found them.

On the 15th of Februray 1564 Galileo Galilie was born in Pisa Tuscany. He was the eldest son of influential musician Vincenzo Galilei. At an early age the family moved to Florence and here he was schooled at a monastery. In 1581 he moved back to his birth town Pisa where he studied medicine at the university of Pisa. Much to his fathers displeasure though, Galileo decided to change his studies from medicine to mathematics and Philosophy, where he excelled. In 1589 he was to become the professor of mathematics at Pisa and in 1592 he moved to a higher salary as the professor of mathematics at the university of Padua. Galileo was an incredible scientist and he worked on a variety of experiments, including gravity and motion. In 1609 though, this is where I think it gets really interesting, Galileo received some news regarding an invention called a spyglass. He was informed that this device using curved glass could

Chapter 6 - A Bit about Moons

make far away objects appear much closer. Without having seen or held a telescope he decided to try and make one. He succeeded. Not only that, after a lot of trial and error he made a much superior model to the one he had heard about. Pretty impressive, don't you think? I'm not sure whether anybody else had thought of pointing the spyglass to the heavens or even if their inferior devices would have worked, but we do know that Galileo did just that. It was believed at the time that the moon was a smooth sphere, probably as a heavenly body should be. Divine and perfect. What Galileo viewed was very different though. He saw with his new optical viewing apparatus a world with mountains and valleys. Great scars adorned the surface of this celestial neighbour.

In 1610, he decided to take a look at the planet Jupiter. Here he made a remarkable discovery. The planet seemed to have four stars surrounding it. It didn't take long for the outstanding astronomer to realise these were actually moons.

These were the first moons, other than our own, ever to be discovered and they are, quite charmingly, called the Galilean moons. They are:

- Io
- Europa
- Ganymede
- Callisto

The nearest moon to Jupiter is Io, it was also the first Galileo discovered. Io was named after a nymph that Zeus transformed into a cow to keep secret his infidelity. Seems a little harsh for poor Io!

Relatively Spacious

With a radius of a little over 1,131 miles (slightly bigger than our moon), it orbits its great celestial master at 262,000 miles.

Io is a world of fire and ice. Active volcanoes adorn the surface, as do mountains taller than Everest. In fact, Io is the most volcanically active world in our solar system. It's also freezing cold with surface temperatures around minus 163 degrees Celsius.

Different chemical silicates on the surface give the moon its splodgy, pizza-like appearance.

Being so close to Jupiter, Io is squeezed and stretched by the giant planet's gravitational effects. This causes tidal heating, and space probes have witnessed volcanic plumes rising up as much as 300 kilometres into space!

We have the Voyager and Pioneer missions to thank for most of what we know about this charming moon.

Io: assets.nasa.gov/image

Next up is Europa, which is similar to the size of our satellite but smaller than Io, in fact, it's the smallest of the Galilean moons.

Chapter 6 - A Bit about Moons

Europa was a beautiful human woman who was kidnapped by Zeus, who appeared to her in the form of a bull. He obviously has a thing about cattle. So taken by this beautiful creature, she climbed on its back and was immediately carried off to the island of Crete where he mated with her under an evergreen tree. Presumably in his human form! They had three sons together, who, upon their death, became the judges of the underworld.

Galileo couldn't have known what we now do today, that Europa is very special.

The surface of Europa is frozen water, and it's incredibly smooth which indicates there is tectonic movement. It has been theorised that the gravitational effect of Jupiter could possibly keep a tidal movement underneath the surface. If this is the case, it would be a very interesting place for astronomers to look for signs of life. Could there be microbial life wriggling around under 15 miles of thick ice in a salty ocean fed by hydrothermal vents? It is possible. Only a future mission can unlock the secrets of Europa.

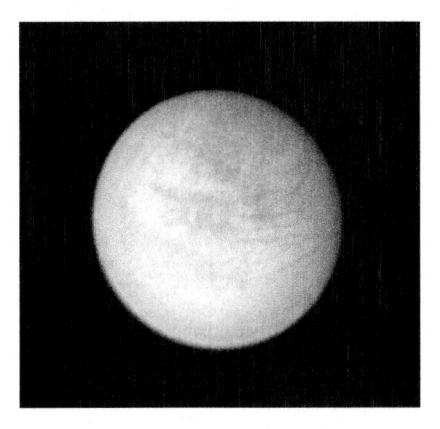

Europa: images-assets.nasa.gov/image/

Ganymede is actually the biggest moon in the solar system. It's bigger than Mercury and only a bit smaller than Mars. It has a radius of over 1,600 miles making it 2.5 times smaller than earth.

It's the only moon to have its own magnetic field, and NASA believes there could be liquid oceans, 60 miles deep, trapped under the 95-mile-thick liquid ice on the surface.

Again, we can thank the space probes for most of what we know about these moons. The Galileo spacecraft skimmed over the moon's surface in 1996, less than 200 miles above, and sent back great close-up images of this king of moons.

In Greek mythology, Ganymede was an attractive Trojan prince. The ever-resourceful Zeus this time transformed himself into an eagle and took off with the prince to become a cup-bearer for the gods and a lover for himself.

What was Ganymede's fate? He was turned into the constellation Aquarius, which signifies the water bearer.

Ganymede: images-assets.nasa.gov/image

Chapter 6 - A Bit about Moons

Callisto is the last of the Galilean moons from Jupiter and my favourite. It is also the most heavily cratered surface in the solar system, which kind of indicates there's not a lot of geological activity going on. One crater is particularly huge, nearly 4,000 kilometres across!

It is a rocky and icy moon, but it is possible there is liquid water deep below the surface. Callisto orbits Jupiter at over a million miles away from the immense planet.

I don't know why, but there's something about Callisto that I like. Perhaps it's the story of how Callisto the nymph was seduced by Zeus but then turned into a bear by Hera, Zeus's wife. Zeus then turned her into a star forming part of the Great Bear star formation! Cute, sort of.

images-assets.nasa.gov/image/PIA00457/PIA00457~orig.jpg

Saturn now has officially 82 moons thanks to observations from the space probes.

One in particular is rather special: Titan.

Titan is special due to the fact it has an atmosphere and is the only moon we know to have one.

It's another large moon, again bigger than Mercury but smaller than Ganymede, but only just.

Discovered in 1655 by the Dutch astronomer Christiaan Huygens, scientists have been excited by Titan, as they now know its main composition is ice and rock.

On January 14th, 2005, ESA's Huygens probe (courtesy of a lift from NASA's Cassini spacecraft) landed on Titan.

There is a YouTube video of the probe descending through the atmosphere and landing on the moon. It's absolutely incredible, and I recommend watching it. You even get to see the shadow of the parachute as it lands! To this date, there is no other spacecraft that has landed this far away from our sun.

Huygens landed on a loose, sandlike surface made up of ice particles. It also found evidence of flowing liquid on the surface and a temperature of minus 170 degrees Celsius. There were many other useful observations taken by Huygens in its final few hours of transmission.

It's strange to think of Huygens, lying there, on that alien world. This metal probe has become part of this enticing moon forever, sinking into its surface and becoming a metallic fossil for the rest of Titans time.

Chapter 6 - A Bit about Moons

Titan: images-assets.nasa.gov/image

These strange remote worlds, held captive by enormous heavenly bodies, in the dark, cold reaches of the solar system is where we look to find life now.

Moons are satellites, and they have had a big part to play in the solar system and have become great planetary companions. Our planet simply would not be the same without one.

Landing on the moon was an exceptional time for space exploration.

In the next chapter, we will look at other incredible scientific feats that we have achieved.

CHAPTER 7

Awesome Space Technology

"When I orbited the earth in a spaceship, I saw for the first time how beautiful our planet is. Mankind, let us preserve and increase this beauty, and not destroy it!"
– Yuri Gagarin, Russian cosmonaut

We have discussed previously the complications just in leaving earth's atmosphere. Imagine then the complications that must arise when planning a trip to a planet 34 million miles away!

When you consider the task, it still amazes me what we have achieved. In my mind, this is what defines us, makes us the best of what we can be, evolution at its greatest; it's what sets us apart in the animal kingdom.

Could a group of hedgehogs or walruses contemplate leaving the earth for a mission to a moon or another planet?

Below are some of the most outstanding achievements of the human race. There have been some blunders along the way, but we are only human after all!

The space race, and probably the Cold War, was launched into international competition when the Soviet Union sent a beach

Chapter 7 - Awesome Space Technology

ball-sized lump of metal into orbit around earth on the 4th of October 1957. Sputnik was a silver-coloured, circular body that carried four long antennas. It was, to put it mildly, very basic, however, it got the attention of the Americans. Plans were being made for their own space programme, as Sputnik raced around our planet every 98 minutes.

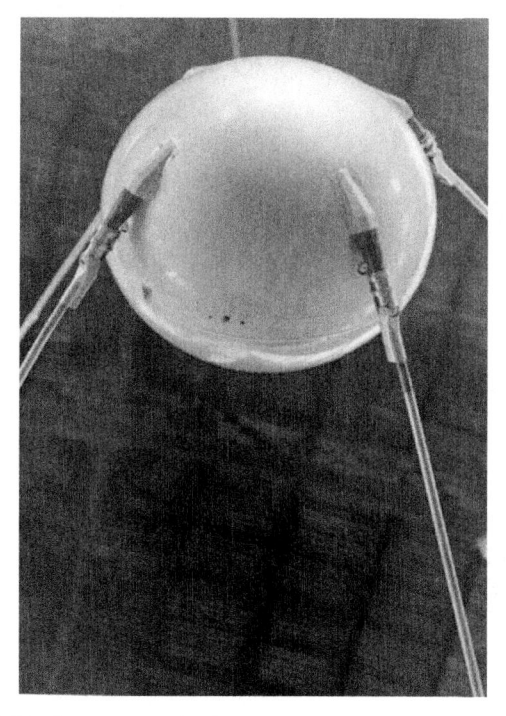

Sputnik Pixabay copyright free images-assets.nasa.gov/image

And so, in 1958, came the birth of an American organisation called NASA. It was formed in reaction to the success of Russia's Sputnik 1. Not to be outdone, NASA was formed, albeit rather hastily!

NASA stands for: National Aeronautics and Space Administration.

It is, in my opinion, one of the greatest organisations ever developed, from its humble beginnings of launching converted ballistic missiles with monkey passengers to the edge of space, to landing men on the moon and sending probes throughout the solar system and beyond.

It hasn't always been easy going, and the ultimate price has been paid more than once.

Seventeen astronauts have perished in the perilous quest for space travel.

In 1967, three men were killed on board the *Apollo 1* spacecraft before it had even left the ground. A fire broke out during a routine procedure, and the three astronauts were burnt to death with no way of escape.

Seven died in 1986, including a schoolteacher, when the space shuttle *Challenger* exploded. After a long investigation, it was found to be an issue with the craft's O-rings, or seals, that had become brittle during colder temperatures.

Seven more were to perish when the shuttle *Columbia* broke apart upon re-entry in Feb 2003.

Sending rockets from earth into space is expensive, troublesome and dangerous. Controlling millions of tons of highly explosive material travelling over 17,000 miles per hour is, well to put it mildly, just incredible. Then to convince a human to strap themselves into a seat on top of one of these enormous fuel containers seems even more incredible.

Chapter 7 - Awesome Space Technology

Let's get this into perspective. Astronauts are pretty much among some of the bravest and most heroic people ever born.

My fascination with space has defined me as a person, and I don't think many days have gone by in my life where I haven't thought about space, or gravity, or time, or perhaps all three.

Would I put myself in a chair on top of millions of tons of combustible oxygen and hydrogen, which I know will be deliberately ignited so I could reach the mesmerising speed required to leave my home planet? Hmm, not for me I am afraid!

I am in awe of these brave adventurers, though.

Have you ever witnessed an astronaut being rescued from his landing craft after being in space a while? It's not a pretty sight. They have to be helped out and carried off, as their legs and arms have ceased to function properly. It's not a dignified, fist-pumping kind of vacating a space capsule that I had imagined. Gravity has played havoc with their senses, particularly their sense of balance, and it can take weeks, or even months, to overcome these effects.

Why do NASA and these astronauts bother?

For science, for knowledge and for answers.

The information NASA gets from space is around 12,000 gigabytes a day. A day!

An average smartphone has between 32 and 64 gigabytes of

storage. Think how many songs and photos and games you can hold in your mobile's memory!

So, man has been sending space probes into space since the 1950s, which, when you think about it, is pretty amazing, considering the Wright brothers only got off the ground, and that was very briefly, in 1903!

The first probes were sent out to the rocky, inner planets in the early sixties, and these were the Mariner missions.

In 1971, Mariner 9 was the first man-made object to go into orbit around another planet. Now that doesn't sound incredibly overwhelming, but imagine the mathematics involved in pulling this off, with not much more technology than a pocket calculator.

This is taken from the NASA Mariner 9 in-depth log as it attempts to orbit Mars:

> **At 00:18 UT Nov. 14, 1971, Mariner 9 ignited its main engine for 915.6 seconds to become the first human-made object to enter orbit around another planet.**
>
> **Initial orbital parameters were about 870 × 11,130 miles (1,398 × 17,916 kilometres) at a 64.3-degree inclination. Another firing on the fourth revolution around Mars refined the orbit to about 870 × 10,650 miles (1,394 × 17,144) at 64.34-degrees inclination.**

Wow! So, they had to ignite those engines at exactly the right time for 915.6 seconds at the parameters above nearly 54

Chapter 7 - Awesome Space Technology

million kilometres away! And what happens if they got it wrong, misjudged it a tiny bit? Gone, forever. The Mariner 9 would have either burnt up into the atmosphere or bounced off the atmosphere into space to be lost forever. Pretty remarkable, hey!

Mariner 9 went on to complete its mission. It took pictures of Olympus Mons, the biggest volcano in the solar system. It also identified Valles Marineris, a vast system of canyons far bigger and deeper than the Grand Canyon on earth. Mariner also sent back images of the two moons we mentioned earlier: Phobos and Deimos.

Mariner 9 is still in orbit around Mars, but it is expected to crash into the planet in 2022.

Mariner 9: image courtesy of Mars.NASA.Gov

I guess it was inevitable that we would land on the red planet. Landing on Mars, though, is not easy. In fact, it's a bit of a nightmare. It's totally riddled with problems.

Mars does have an atmosphere, but it is a very thin one, a hundred times thinner than earth's.

Heavy heat shields are still required upon entry, but the atmosphere is too thin to slow you down. You will need rockets, which will require fuel, to slow you down, but these also make you heavier! Parachutes don't work as well on Mars due to this meagre atmosphere.

Once you have overcome these issues, then you need to land. Luckily, scientists at NASA have come up with some very interesting ways of doing this, and I think Pathfinder is my favourite.

The end of the Cold War, unfortunately, had a negative effect on NASA's space budget. Cuts had to be made, so the pressure was on to be efficient and cost-effective. NASA's engineers were up to the challenge though.

Mars Pathfinder was launched December 4th, 1996 and landed on Mars's Ares Vallis on July 4th, 1997. It had travelled 120 million miles to get there. The landing process was impressive and unique but, above all, cost-effective.

First, it used parachutes to help slow it down a bit, once it was clear of the thin, Martian atmosphere, then it deployed huge airbags to cushion its impact at 40 miles per hour on the surface. It bounced back up into the Martian sky 16 times before landing safely. Pathfinder and its rover, Sojourner, were a complete success.

Little interesting fact: if you have ever seen the film *The Martian*,

Chapter 7 - Awesome Space Technology

well, Pathfinder is the buried rover that Matt Damon goes off to find so he can make contact with earth!

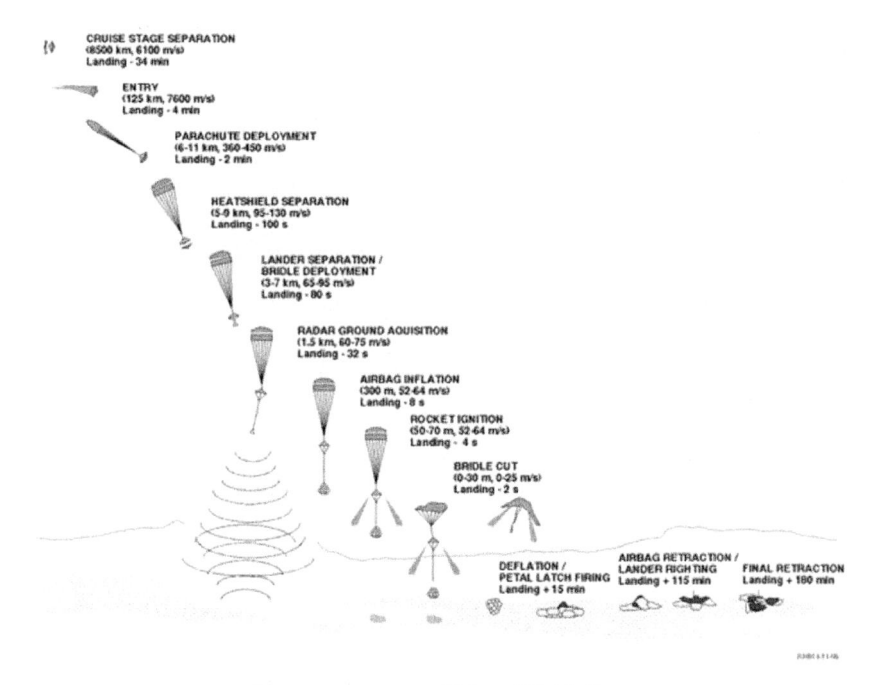

```
CRUISE STAGE SEPARATION
(8500 km, 6100 m/s)
Landing - 34 min

ENTRY
(125 km, 7600 m/s)
Landing - 4 min

PARACHUTE DEPLOYMENT
(6-11 km, 360-450 m/s)
Landing - 2 min

HEATSHIELD SEPARATION
(5-9 km, 95-130 m/s)
Landing - 100 s

LANDER SEPARATION /
BRIDLE DEPLOYMENT
(3-7 km, 65-95 m/s)
Landing - 80 s

RADAR GROUND AQUISITION
(1.5 km, 60-75 m/s)
Landing - 32 s

AIRBAG INFLATION
(300 m, 52-64 m/s)
Landing - 8 s

ROCKET IGNITION
(50-70 m, 52-64 m/s)
Landing - 4 s

BRIDLE CUT
(0-30 m, 0-25 m/s)
Landing - 2 s

DEFLATION /           AIRBAG RETRACTION /
PETAL LATCH FIRING    LANDER RIGHTING      FINAL RETRACTION
Landing + 15 min      Landing + 115 min    Landing + 180 min
```

Image courtesy of Mars.NASA.Gov

After Pathfinder came Spirit and Opportunity. Launched in 2003, they were much bigger rovers than their predecessors and were examining the planet's geology and atmosphere. These rovers were the first to identify that the red planet once had rivers and even oceans.

I named this chapter amazing space technology, and it is so befitting to the Opportunity rover. Although it was originally designed to work for 90 days, JPL (Jet Propulsion Laboratory) had engineered a rover so well it went on to operate for 14 years! It cost 400 million dollars for Opportunity. I wonder what the cost would be today for a 14-year, operating rover on a planet

53 million miles away?

On August the 5th 2012, NASA successfully landed the Curiosity rover on Mars, the biggest and most advanced rover yet. Its purpose? To look for signs of life in the rocks of ancient Mars.

Curiosity is still roaming around on Mars, and if you click on to NASA's Mars Exploration Program, you can see where Curiosity is at that exact time!

Mars really seems to have captured the human imagination. In February 2021, there are three missions that are due to land on Mars!

On February the 9th 2021, the United Arab Emirates hope to insert a probe into orbit around Mars called... er, funnily enough "Hope".

It's the first interplanetary probe from the Arab nations, and its mission is to get a good understanding of the Martian atmosphere.

A day later, China's Tainwen-1 probe is set to go into orbit around Mars. It will be China's first independent Mars assignment and, if all goes well, it will deploy a small rover on the planet!

On February the 18th, NASA intends to land its largest and most advanced rover yet on the red planet; Perseverance.

Strapped to the underside of Perseverance is a small helicopter drone called Ingenuity!

Chapter 7 - Awesome Space Technology

So, Mars is going to be a busy place soon. Let's hope all three probes make it safely. As you now know, landing safely on Mars is a very tricky task. In fact, it has been estimated nearly half of all Mars missions have ended unsuccessfully.

Studying the inner, rocky planets has been a real testament to scientific development despite its many hurdles and staggering distance.

So, when the guys at NASA wanted to study the outer gas and ice giants this was a really big ask. There have now been many probes sent out to the outer solar system, but I think some of mankind's greatest achievements are the Voyager missions, and I feel compelled to tell their story here.

When you're looking to travel to the outer solar system, you need all the help you can get. In the middle of the 1960s, in a bizarre planetary coincidence, help was at hand.

Gary Flandro, a Caltech graduate student, was checking to see if there was an alignment of the gas giants, at some point in the future, so they could be used as multiple gravitational slingshots. Its purpose would be to send a spacecraft from one celestial body to another.

A slingshot is where a spacecraft goes into orbit around a planet, to gather more speed and momentum, before it then speeds off to its next destination.

Incredibly, he found there was!

There would be a very rare alignment of Jupiter, Saturn,

Neptune, Uranus and Pluto in the mid eighties.

It would save nearly 20 years of space travel, to use these slingshots, rather than travelling directly to the planets.

If something was meant to be it was the Voyager missions. It does seem a little like fate that, just when we had the technology and were looking for the alignment, we found it. We also found it with enough time to build the two Voyager probes. This alignment of the planets won't happen again for nearly 200 years.

Carl Sagan and a committee were put in charge of deciding on a message to put on board Voyager.

They decided on a golden record adorned with symbols, the human body, male and female, the position of our location in the Milky Way and all kinds of sounds including music. There would also be attached a needle and instructions on how to play the record! It's a greetings message to any beings who may come across the probes in the distant future. These messages weren't considered by all to be a great idea, though. Some thought it wasn't wise to give away our location to potentially hostile aliens!

Chapter 7 - Awesome Space Technology

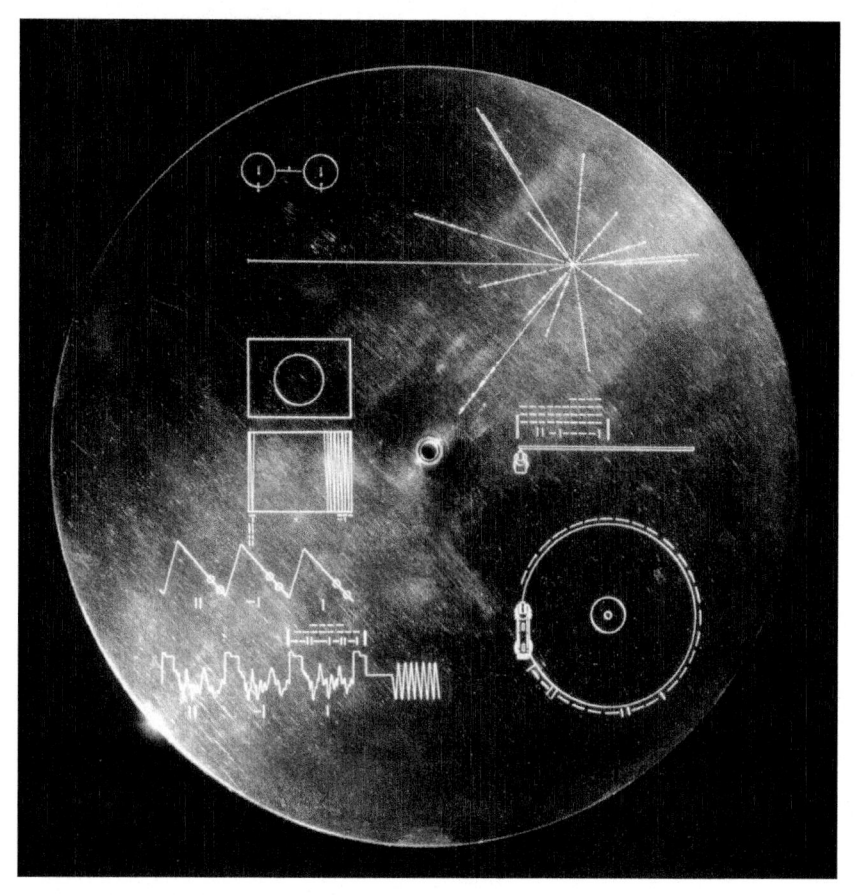

Image credit NASA

Launched in 1977, both Voyager 1 and Voyager 2 have been an absolute success, beaming back amazing images of the outer planets and even discovering new moons. Incredibly, and beyond expectations, they are still sending back data now even though they have left our solar system.

As I write this, Voyager 1 travelling at 35,000 miles per hour and is 22 billion kilometres away, and Voyager 2 is some 18 billion kilometres from earth.

Relatively Spacious

The next destination for Voyager 1 is the constellation Ophiuchus, where it will come reasonably close to a small star. This will take over 38,000 years, however. Voyager 2 will come close to a star in the Andromeda constellation in around 40,000 years.

What then? Well, they will just keep going, hurtling through space, orbiting the centre of the Milky Way for billions of years. It's possible, in fact probable, they will still be travelling after our sun has burnt out and our solar system is no more. Only a collision will stop the Voyagers now.

Not bad for a couple of metal probes hastily put together to make the most of a planetary alignment!

I remember being at school when I first heard that NASA planned to send a telescope out into space. I imagined a big sort of floating observatory, with huge metallic domes riddled with large, rounded, shiny rivets and huge, rectangular openings with large, bronze telescopes pointing out to the cosmos, a place where astronauts could visit to view the stars. Of course, it turned out to be nothing like that, but I think it's still one of the coolest things we have done as a species.

The idea had actually been floating around for decades. Astrophysicist Lyman Spitzer Junior had written a paper, back in the forties, about an observatory in space that would be free of earth's atmosphere. It took a while, but in 1977 it got approval.

The huge space telescope was to be named after Edwin Hubble.

Hubble was a remarkable character, and it's easy to see why

Chapter 7 - Awesome Space Technology

they chose to name the world's greatest telescope after him.

Born in November 1889, Hubble was to become a brilliant American astronomer. He was tall and athletic, an excellent athlete in both track and field and an accomplished boxer. He was so good looking he was likened to Adonis by his friends. It all seemed just to good to be true. It actually wasn't!

During his early school years, he was recognised as exceptionally bright. He would go on to break the state high jump record.

In the late summer of 1906, Hubble enrolled in the University of Chicago.

He left Chicago with a degree in maths and astronomy. In 1909, he was to obtain a scholarship to study at Oxford University in England. It was a highly sought-after scholarship and only one student from each state was chosen.

He was to receive second class honours at Oxford and wanted to stay and complete a bachelor's degree. Instead, though, he was to move back to America, to look after his family, after hearing of the passing of his father. In Kentucky, where his family resided, he took a teaching position, but it wasn't long before his longing for astronomy became too much to ignore. He received a position at Yerkes Observatory in Wisconsin, and this was where he started his observations and collecting data on nebulae. At this time, nebulae were just seen as bright, fuzzy clouds.

Hubble made some interesting breakthroughs in this field which caught the attention of George Hale who was the director of the Mount Wilson Observatory in California.

The Mount Wilson Observatory could have its own book, it was a titanic struggle to get off the ground, and it was fraught with problems. There are even poems that have been dedicated to it.

In 1917, Hale offered Hubble a position. Unfortunately, due to the First World War, this appointment would have to wait. In 1919, though, Hubble finally made his way to California and the great observatory.

Here, with the latest technology and a 100-inch telescope, he correctly observed that the Milky Way was just one of millions, or billions, of galaxies in the universe.

He then went on to observe that galaxies were moving away from one another which led to the realisation that the universe is expanding.

Hubble was involved in both world wars and was regarded extremely highly, some regarding him as the best scientist since Galileo.

Hubble didn't win a Nobel Prize for his discoveries, though, which is really rather strange.

So, I guess it's fitting and fair that something so amazing as this space telescope was named after him.

As in most cases, when you attempt something in space it's a lot easier said than done. There were tremendous hurdles to overcome, and a lot of science and mathematics and planning to

Chapter 7 - Awesome Space Technology

launch such a delicate, but enormous, viewing device into space.

There is an interesting book called *The Hubble Space Telescope*, by a chap called David J Shayler, which recounts the Hubble Space Telescope from concept to success.

It's an excellent book and very compelling reading, especially the narrative describing the deployment of the telescope when in space. I highly recommend reading it if you would like to know more about this.

The space telescope was about to suffer a string of delays, though.

Originally destined for take-off in the mid eighties, it was put back due to the terrible *Challenger* tragedy where all six astronauts were killed just after lift-off. The aftermath of the *Challenger* tragedy put NASA's payload back years, and it wasn't until April 24th, 1990 that the space shuttle *Discovery* launched from Kennedy Space Center with the world's most expensive telescope safely stored on board, all 43 feet of it!

There were five astronauts charged with delivering Hubble to space:

- Loren Shriver
- Charles Bolden
- Kathryn Sullivan
- Bruce McCandless II
- Steve Hawley

It was, as I said, a very difficult mission with many complications.

There were issues with the solar panels not deploying properly and problems unloading the enormous telescope from *Discovery* with a giant mechanical arm. These were just a few of the hurdles they had to overcome.

Eventually, however, at a distance of 330 miles from earth, Hubble was put into orbit around the earth travelling at around 17,000 miles per hour.

NASA and the world waited with anticipation for the first images.

Disastrously, though, there seemed to be a problem.

The images beamed back by Hubble appeared blurred or out of focus. It was a devastating blow.

They soon discovered the main focusing mirror had been polished to the wrong specification.

The media were ruthless, claiming it was a 1.5 billion dollar disaster.

In 1993, though, during the first Hubble servicing mission, NASA installed "COSTAR" a collection of small, movable mirrors which acted like a pair of glasses to the giant telescope. Hubble was soon sending back incredible, mind-blowing images of the cosmos.

It can look so deeply back in time, it has helped us date the universe to around 13.8 billion years old.

Chapter 7 - Awesome Space Technology

Hubble is soon to be replaced by a new telescope, the James Webb Space Telescope. This is due to be launched in 2021 and will go into orbit 1.5 million kilometres from earth, way past the moon. It will be looking for the first galaxies formed by the Big Bang and how they look now. This telescope will look in the infrared spectrum, and the lens will be eight times larger than the Hubble telescope.

I guess sending a telescope out in space was cool, how about an international space station? Well, in 1998, that's exactly what they did. Well actually, not exactly, if you remember from a previous chapter about escape velocity, there is no way we could send a space station into space. It weighs over 400 tons, so there is just no way we could escape earth's gravity carrying something so big.

So, it would have to be made in parts and assembled in space.

In one of the biggest world co-operations and joint missions ever, a partnership between the ESA (European Space Agency), NASA, JAXA (Japan), CSA (Canada) and Roscosmos (Russia) was put together to plan, manufacture and build the space station.

So, since 1998, we have been building the space station in space. In fact, since 2006, it has been permanently occupied by one country or another.

Orbiting earth at five miles per second, at 250 miles above us, it's one of the brightest lights in the sky. It's about the size of a football pitch, and I have had the pleasure of watching it pass overhead numerous times.

Relatively Spacious

If you want to see one of mankind's greatest achievements from your garden or bedroom window, go on to NASA's Spot the Station app, and they will text you with the best times to see it or when it will be due to be passing over your area.

To date, it is the most expensive object ever built at 120 billion dollars. I should imagine, unless we do a manned mission to Mars, or put a station on the moon, it will probably stay that way for a long while.

The International Space Station has benefitted us here on earth in many ways. Not to mention the stunning photos it beams back to earth of our planet.

Water purification and filter arrangements, designed for space travel aboard the ISS, have had enormous benefits to some countries.

Latest technology in eye surgery has developed on board the station, as well as technology to use a bionic arm to perform surgery on previously inoperable tumours. There is also groundbreaking advancements in medicine, vaccines and vital progress with work on muscle loss and osteoporosis.

It is, to some extent, one of mankind's greatest achievements. I guess what I like about it most is that it's a collaboration. All who work aboard this vessel have one thing in common; the quest for knowledge.

Chapter 7 - Awesome Space Technology

The Hubble Space Telescope: NASA stock images

International Space Station: NASA stock images

What would you do if you turned on your TV to see a news broadcast about a giant asteroid on its way to earth! Panic probably and rightly so, I would too.

I have mentioned that there have been asteroids in the past that have nearly ended all life on earth. Fortunately, though, you and I live in a time that could potentially deal with such a threat. If any asteroid of considerable size (over a mile wide) had collided with our planet anytime in the last few hundred thousand years, up until about 40 or 50 years ago, it's most probable that humans would have had the unfortunate privilege of seeing their own extinction literally bearing down upon them.

Most asteroids are travelling at such high speeds that, if they are large enough to break through our protective atmosphere, they could cause significant and devastating damage.

Large asteroid impacts are quite rare, but smaller ones are probably happening all the time. Anything around a mile wide, though, is what scientists would call extinction-level asteroids, and there are plenty of them out there!

Basically, they're life-ending rocks smashing into earth at over 35,000 miles per hour.

In 2013, a meteor exploded above the Russian city Chelyabinsk. It exploded with the force of about 30 nuclear bombs. Luckily, it exploded high in the atmosphere, so damage was limited, but it still caused injuries to 1,400 people. These kinds of strikes probably happen quite frequently! It's just they normally strike in remote places, so we don't notice.

In 1908, on the 30th of June, a massive explosion occurred in Siberia near the Tunguska River. The meteor was believed to be about 100 metres wide and it flattened 80 million trees.

Again, luckily for us, it impacted in a very remote area.

Chapter 7 - Awesome Space Technology

The problem we have is making sure we find these potential collisions before they happen. Scientists and governments now take this very seriously and are on the lookout all the time, helped out by the watchful eyes of some amateur astronomers as well.

Once something is identified as a potential threat, it's classed as a NEO (near-earth object) or NEA (near-earth asteroid).

At first, it was generally considered that the best way to deal with these things would be to send a nuclear warhead into space and blow this speeding mass of, well, whatever it's made up of, into lots of smaller whatevers.

However, these smaller bits of asteroid could possibly just end up end falling to earth but would now be covered in harmful radiation.

Scientists, luckily, have come up with other techniques.

On the 12th of February 2001, after a journey of nearly five years, a small space probe from NASA landed on a strange, potato-shaped asteroid. The spacecraft, the "Near-Earth Asteroid Rendezvous" (NEAR) Shoemaker, was on mission to study this strange, 20-mile-long celestial body we had named Eros.

It was the first time humans had landed a spacecraft on an asteroid. It was a marvellous achievement and credit should be given to all involved.

Why did NASA want to land on this irregular rock though?

Well, Eros was actually identified back in 1898, and it had been classified as a NEA (near-earth asteroid). In fact, it was the first of these to be discovered. Eros is in an orbit around the sun that throws it out towards Mars then back again towards earth.

Although there was a lot of geological interest in Eros, it was also a chance to see up close one of these NEAs and learn more about them. This particular asteroid is one of many that come within a few million miles of earth.

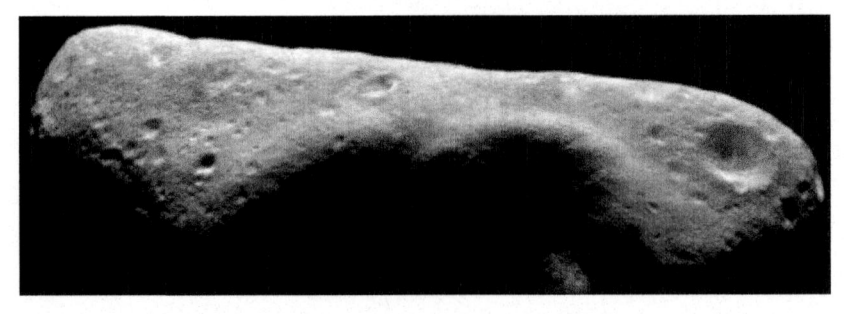

A sidelong view of asteroid 433 Eros: NEAR Shoemaker.

Scientists need to know these things so they can then judge the best ways to deal with these orbiting missiles should they turn out to have an orbit that might one day collide with earth. To understand Eros might help them get an understanding of other NEAs. What are they made of? Are they solid? If we heated them would they melt a bit?

If we can land a spacecraft on an asteroid, potentially then we could land thrusters on one to force its trajectory slightly or perhaps even just smash a probe into one.

It would only need to knock it off course by a few millimetres as, over time, this would make a huge difference. This could even

Chapter 7 - Awesome Space Technology

be done just by sending a probe into orbit around an asteroid and letting gravity do its stuff, nudging it ever so slightly, where again, over time, it could miss by millions of miles.

These are all solutions, of course, if we happen to find an asteroid with lots of time to spare, so we can make detailed plans and build a craft to carry out this intricate mission.

Although NEAR Shoemaker landed on Eros successfully, there was a lot that could have gone wrong; I guess there always is when you are trying to land something on a rock travelling at around 50,000 miles an hour, 196 million miles from earth!

Plenty of times we have had near misses of asteroids, or meteors, that we didn't know about until they had fortunately, and safely, hurtled past.

It is a very real and serious problem. Fortunately, there are astronomers, professional and amateur, surveying the solar system and cataloguing these very dangerous, potentially extinction-level, rocky giants.

Life is precious, and it's worth protecting.

CHAPTER 8

Life

"Nothing in life is to be feared, it is only to be understood. Now is the time to understand more, so that we may fear less."
— **Marie Curie**

What makes humans, human?

I have a lovely, blue Staffordshire bull terrier, funnily enough, called Blue! Now I know Blue has a brain. In fact, she is capable of some pretty clever stuff. She knows her name, she knows where she sleeps, and she can make her way to her food bowl without any trouble at all. She can differentiate between a human and a squirrel; she loves humans, hates squirrels. She can pick out words that will make her ears prick up. In fact, she's actually capable of some actions that are, well, beyond human capabilities. If I'm late home, she won't bark at the sound of me coming through the back door, even though she can't see me, but she would at a stranger walking down the drive she couldn't see. This is clearly an ability or sense I do not possess. If, however, I asked Blue to help me put up the Christmas tree, or peel me a banana, then we would have a problem. Some of this would be language and some of this would be that she is not physically capable of it, but mostly it's because she couldn't comprehend the action.

Chapter 8 - Life

We know we are related to dogs at some point in the past. What happened, then, to make us so completely different? Why is it I spend my days thinking about space and time, whilst Blue is content just to have a walk and be fed? Blue doesn't know about stars, or galaxies, or space-time.

There have been theories put forward that human brains really began evolving once we had mastered fire. With fire, we were able to cook our food. Cooked food meant that we didn't need to waste endless hours trying to digest it. So, more hours to think, more time to look at the stars and contemplate our existence.

Bigger brains meant better tools to hunt with, which again meant less time hunting and more time thinking. One of mankind's greatest achievements has been the ability to sew. Sewing meant a better defence against the cold and better camouflage. Clothing made it possible to be out at night looking up to the heavens, looking for answers to our existence and, of course, to survive different extremes of temperature like ice ages.

Before we go too far into human evolution, I think we should go back to the very beginning. It would be easy to concentrate on human life, but that would be missing out on 99.99 per cent of all the other life on earth. Again, you will see we seem to have been incredibly fortunate to be here at all.

As far as we know, life has only happened once. Life isn't continually springing out from primordial soups or giving animation to pebbles or rocks.

We know, from dated rocks about 3.5 billion years ago, life

possibly started in the oceans, or rocky pools, conceivably heated by volcanoes or thermal vents. This was very basic, single-cell life. Everything alive today, or what we consider alive, is descended from that first being, creature, or blob of algae.

This single-celled little bag of life seemed pretty content, as it stayed that way for two to three billion years before evolving into anything, well, anything much more than a little bag of life.

Now, when you think about it, it's quite a lot to take in, and it does kind of make you think about the beginning of the universe all over again. First there was nothing, and then there was something! First there was no life, then an abundance of life. How?

There is a great recorded argument between two Frenchmen in the mid 1800s; Louis Pasteur, who we can thank for his work developing vaccines for cholera, smallpox and others, and Félix Archimède Pouchet, a French naturalist.

Pouchet believed that life came from inanimate objects, even from the air.

This was, of course, pretty much the thinking back then in the 1800s. Pasteur, though, had reason to believe this wasn't the case.

Doing work on germs and bacteria, he could see that life was born from previous life. He saw that in sterile conditions life didn't arise.

Chapter 8 - Life

It's very easy to see how confusing this must have been for past generations. Looking at dead, rotting meat, it would have seemed that maggots just appeared from nowhere to feast on the available meat. Of course, we now know that maggots come from flies' eggs that would have been laid on the flesh.

This spontaneous generation theory, as it was called, really started with the Greeks and Romans, who looked for other explanations of life other than a divine one.

Fish, it was thought by one Roman philosopher, Anaximander of Miletus, were brought into being by warmed up water and mud or earth. These fish would eventually seed human life.

It's a good idea, but he doesn't go on to say how this occurs.

Anyway, back to Pasteur and Pouchet!

The French Academy of Sciences declared a prize to whoever could prove, through scientific process, the evidence of either argument.

It was an unfriendly and bitter battle between the two scientists, and both parties claimed that the experiments of the other were conducted poorly. Eventually, however, over a period of time, Pasteur's work gained recognition.

Pouchet ultimately pulled out from the competition, and Pasteur was awarded the prize. It was a good result for scientific experiment and Pasteur, who was a mighty scientist. He continued working despite suffering a stroke which left him partly paralysed.

There is a big problem though, as Pouchet noted! If life doesn't arrive from nonliving matter, how did it get started in the first place?

Well, life did start, otherwise we wouldn't be here now; unless we are a computer-simulated game for a more advanced race.

This actually has been theorised by some scientists as a very real possibility. However, we would then have to think about who gave life to them?

Another hypothesis is the panspermia theory; that life drifted down from space. Good theory, we know all the building blocks for life are out there, but again it doesn't really answer the question of how it started.

The answer, of course... is we don't know. Nobody does.

Scientists have tried to imitate early earthlike conditions in labs with chemicals, water and electric currents to simulate lightning strikes but, although they have had some success in forming certain chemical reactions, they have not been able to achieve anything we would call life.

Life is really quite curious. You see, we are all made up of atoms. Atoms are incredibly tiny things. It would take about a million of them to make up the thickness of a page of this book. You only need a few dozen different atoms, and you can make whatever you like, a killer whale, or an ironing board, or an albatross wing, or a brick, in fact, anything or everything.

These atoms are not alive though. If you start to join them

together, you start to get things. Join a few together, though, in the right way, and you can create elements, gases, or liquids. (Water, for instance, is two hydrogen atoms and one oxygen atom H2O.)

Join atoms together in a certain way and you could make you. But this "you" wouldn't be alive, just a bag of chemicals and gases and liquids. I'm presuming, of course, you think of yourself to be more than this. Of course you do, and so do I. So where did this spark of life come from?

Well, life started somehow and around 570 million years ago it started to transform from single-celled, little beings that had existed for three billion years to more complex life. Then about 540 million years ago, life got big.

You may have heard of it or remember it from school, it's called the "Cambrian explosion".

What sparked the Cambrian explosion? Probably the amount of oxygen being produced by the algae and bacteria in the ocean.

The most notable thing about this era is that many of these creatures had acquired hard body parts or external skeletons.

These hard body parts could act like scaffolding and meant that creatures could grow much bigger than previous species.

This was the era of the trilobite. With thousands of different species, they were incredibly numerous and with hard, segmented, plated bodies, trilobites had armour to protect them against predators.

Life, at this point, was still only in the oceans, but these now busy and habitable bodies of water at some stage, became very unstable as the earth cooled over a few million years.

This could have caused the mass extinction which wiped out nearly half of all life on earth. The trilobites, however, managed to dodge this annihilation!

Fast forward to around 465 million years ago, and we begin to find evidence of the first plants in the fossil records. Then 400 million years ago, signs of insects started to appear.

Approximately 385 million years ago, we can see indications of the first trees.

There are signs that early reptiles were walking on land 300 million years ago in the Permian era.

Despite surviving some catastrophic global extinctions, the trilobites couldn't hold out against earth's biggest extinction though, the one that ended the Permian era about 250 million years ago. It has been estimated that over 70 per cent of all land creatures died out and a staggering 95 per cent of sea creatures were to perish.

What could have caused such global devastation?

The best theories now are that it was caused by global warming due to catastrophic volcanic eruptions. This led to the oceans losing as much as 80 per cent of their oxygen.

Chapter 8 - Life

It may have been this mass extinction that led to the rise of new types of creatures in the fossil record, the first dinosaurs and the first warm-bloodied mammals.

Somewhere around this time, in space, a large rock was being nudged out of its place in the asteroid belt which pushed it into a much closer orbit of the sun's.

Its destination, in a hundred million years: the Yucatán Peninsula, Mexico, Central America, earth. We will come back to this shortly.

The dinosaurs are one of earth's success stories. They reigned for around 165 million years, which is pretty impressive when you think we have managed to cause ecological species-ending conditions on this planet in a hundred years or so!

Dinosaurs were magnificent creatures and capable of growing to extraordinary sizes. In 1986, in Argentina, archaeologists uncovered a huge dinosaur which turned out to be 120-foot long and would have weighed a staggering 100 tons. They named it Argentinosaurus, of course.

The biggest carnivore was a dinosaur called Spinosaurus which outweighed T. rex by a couple of tons and stood seven metres tall and up to 15 metres long!

It wasn't all plain sailing for the dinosaurs, though. There had been extinctions for some subspecies previously, but generally, as a species, 165 million years was pretty impressive. But then all of a sudden, they just disappear from the fossil records. Around 65 million years ago, something happened that ended the mighty reign of the dinosaurs, and it ended their reign swiftly.

Hurtling through space, 65 million years ago, was an asteroid, over six miles wide and travelling at over 40,000 miles an hour. It was the asteroid we referred to earlier and, although scientists are not 100 per cent sure, it's probable that it came from the asteroid belt nudged to an inner orbit by a collision 100 million years before it struck earth.

This silent, cosmic assassin had travelled billions of miles on its journey, and it punched through our atmosphere at astonishing speed above the Yucatán Peninsula near the modern-day coast of Mexico, impacting with the force of a million nuclear bombs. The asteroid would have vaporised significantly before it hit the earth, but the crater it left is 60 miles wide and 18 miles deep. The immediate effects would have been deadly, cooking any creature alive as far away as the impact was visible. Forest fires, earthquakes and tsunamis soon followed. The deadliest effect, though, was the blasted material that ended up in the atmosphere, blanketing out the sun and causing a global winter for years, killing off the dinosaurs and many other species as well; anything big, basically. It was a very bad day for life on this planet.

This wasn't the first major asteroid impact on earth. In fact, there are at least five or six larger impact craters on earth. Nor was it the last.

The dinosaur-ending asteroid was really bad news for the dinosaurs but incredibly fortunate for another group of animals! With the dinosaurs out of the way, there would be a new dominant force on the planet.

It was the time of the mammals.

Mammals had been around for quite a long time, probably living alongside the dinosaurs for a few million years but, in all likelihood due to the threat of being eaten, stayed relatively small, so not much bigger than a modern-day pet cat.

Whatever happened during the downfall of the dinosaurs, we do know one thing for sure, that one line of mammals survived, and all modern humans are related to that line of mammals.

Whereas reptiles and birds lay eggs, mammals would give birth to live young, using a placenta in the womb to incubate and mature. Remarkabley and cleverly, the placenta is isolated from the mother's immune system to stop her antibodies from attacking the father's genes. There are a few exceptions, but this is generally the case.

About 55 million years ago, there appears to have been a noticeable increase in temperature which encouraged super growth for trees and plants around the world. Whether coincidental or not, it is around this era that we find the first primate fossils in East Africa.

Roughly, about 40 million years ago, the shifting tectonic plates between India and Eurasia piled into each other and created what we today call the Himalayas.

This had severe consequences for earth's weather patterns and may have been responsible for a string of ice ages that were to follow.

Whatever the reason, ice ages seemed to have a big effect on mammals, where I mean they got bigger, some of them much bigger!

The Columbian mammoth reached 12-14 feet high, and the ground sloth grew up to 20 feet tall or long and weighed in at four tons! A sloth at four tons!

The cave hyena was twice the size of a modern-day hyena. And of course, there was the smilodon, or sabre-toothed tiger, a huge beast with enormous eight-inch fangs.

These are just a few of the enormous mammals that roamed the earth some 2.5 million years ago. These are all, of course, extinct now along with many numerous others.

It's difficult not to associate their disappearance with the appearance of a new subspecies around 300,000 years ago, hominoids.

Big was probably an advantage against the cold and predators until groups of organised hominoids arrived. Big then became a disadvantage, meaning easier to hunt or, unfortunately for them, less easy to hide.

The oldest recorded fossil of Homo sapiens is in Ethiopia, dating back 195,000 years ago. What drove this species of primate out from the trees and on to the plains?

Well, it's a very good question, and probably the answer lies in

Chapter 8 - Life

the diverse weather sequences driven by the formation of the East African Rift Valley. Again, this was plate tectonics at work.

Hospitable, lush forests suddenly became dry and arid. Forests became plains, perhaps forcing some primates from the disappearing forests on to the plains.

It gets a little confusing here, because I thought that Homo sapiens evolved from Neanderthals, I'm sure that is what I was taught in school, but of course I could be wrong.

It turns out, though, that we actually share some DNA with Neanderthal man, and we lived alongside them and mated with them up until about 30,000 years ago. Unfortunately, again Homo sapiens were probably responsible for their demise. Some of Neanderthal DNA and genes are still found in humans today. There were in fact several types, or species, of hominoid that were around at the same time as Homo sapiens.

If you wanted evidence of how far evolved Homo sapiens had come, around the time the last Neanderthals disappeared, the best evidence would be cave paintings.

This shows a massive jump in intelligence and free thinking. In fact, more than that, you can clearly see some of these cave paintings are clearly abstract, showing great imagination and personal perspective.

In 1994, in the middle of December, Jean-Marie Chauvet, the Ministry of Culture park ranger, and two friends were checking the possibility of some caves which, they had been told by previous explorers a few months prior, could be found near the bank of the river Ardèche in southern France.

Amazingly, they found what they were looking for and a lot more.

At first, they found some caves where there was evidence of some painting but, eventually, they came across what can only be described as a prehistoric art gallery. The cave had been sealed off for at least 20,000 years but had been used numerous times prior to that. The oldest paintings date at least 30,000 years old.

There are images of lions, rhinos, bears, horses and reindeer. The artists had even used some of the natural rocks as parts of the bodies to give the paintings more depth. They are incredible. In my opinion, it's the first real evidence of human thinking.

The Chauvet cave is closed off to the general public now to protect it, but they have a website, and I really would encourage anyone to take a look at these incredible paintings. Remember, these were done with charcoal and plants and no paintbrushes!

Chauvet cave painting courtesy of the bradshaw foundtaion

Chapter 8 - Life

We are not entirely sure how agriculture and farming started or exactly when. Common estimations are around 12,000 years ago, and it's highly likely it started in different locations around a similar time.

Until this time, humans were on the move. We had made it to every continent, thanks to ice age land bridges for some locations like Australia and the Americas.

So why did we settle down and start farming?

Well, farming had its advantages and disadvantages. Being a hunter-gatherer on the move was a difficult and hard life for a family, especially with children, following the seasons and the herds of prey. Farming gave them a more stable base and more consistent availability of food; they would, of course, still forage and hunt, but they would have animals to domesticate and crops to harvest. Children were easier to look after and didn't need a constant vigil.

With this easier way of life, the farmers could have more children, more help on the farm, but then this resulted in more mouths to feed. This of course is only my theory, but you could see how quickly small farms could turn into villages and larger settlements over a relatively short period of time.

With the new introduction of villages, and then cities, the human population exploded from probably a few million to 250 million.

Mankind hadn't stopped his killing spree yet either. Now there were lots of other reasons to kill. Religion and land soon became an excuse to rage war on one another.

Whomsoever you are, you can be sure one of your ancestors was a murderer, or rapist, or both.

Well, on a cheerier note, that was an extremely condensed timeline of life on earth. As you can see though, it didn't take life long to get a foothold but, when it did, it just stayed that way for a few billion years.

It's highly possible we are still evolving. Someday, in the distant future, we may look very different to the way we look today.

Life appears and then goes extinct; it seems to be the nature of things. But it also seems to be the nature that somehow, somewhere life holds on and that, of course, does seem encouraging. But sometimes it seems that life doesn't seem to want much.

Bill Bryson, in his brief history of everything, notes:

> Consider the Lichen. Lichens are just about the hardiest visible organisms on earth, but among the least ambitious. They will grow happily in a sunny churchyard, but they particularly thrive in environments where no other organisms would go – on blowy mountain tops and arctic wastes, wherever there is little but rock and rain and cold, and almost no competition.
>
> Life in short, just wants to be-and here's an interesting point- for the most part it doesn't want to be much.
> (Bill Bryson, *A Short History of Nearly Everything*)

This lichen is, to all intents and purposes, alive. It ticks all the requirements for being alive. It doesn't look forward to a showing at the cinema or a trip to the shops, it doesn't get pollinated by lovely insects, it is just there taking any substance it can from a very cold rock in a very cold place... Living, of sorts, I suppose!

Human life isn't even a blink of an eye in the timeline of the earth, and yet we have changed the planet in such a short period of time.

We have created great sprawling cities, redirected great rivers and moved unfathomable amounts of earth. And yet, from space, you wouldn't know we existed at all.

I like to think that, with the help of science and the great scientists that are out there, we could help save our planet from what seems like its impending doom that we have created with our throwaway society.

I'm sure we are capable, despite our murderous past. I believe most people now are inherently good and that we have evolved to understand suffering and joy.

I read an article recently about a guy who made a prosthetic limb for an elephant that had trodden on a mine.

These acts are human; it shows incredible empathy and kindness.

The elephant couldn't return the debt, even it wanted to.

We alone have the ability to change the world, and science is our strongest weapon.

Unless, perhaps, we had an ally, from another world, which could show us the way and help us to be better.

CHAPTER 9

Alien Civilisations

"The only thing that scares me more than space aliens is the idea that there aren't any space aliens. We can't be the best that creation has to offer. I pray we're not all there is. If so, we're in big trouble."
– **Ellen DeGeneres**

Do I believe in alien life? Yes, 100 per cent. Do I believe in alien life visiting this planet and beaming up lumberjacks from a remote road in the middle of the night? No, 100 per cent no.

Aliens make great movies, and I am a real fan of most sci-fi books and films, but I think a reality check is normally in place when I really think about these things.

As a child growing up, I read a lot of sci-fi books and comics and worried about alien invasion probably more than anything else. I read *The War of the Worlds* and found it terrifying.

The thing is, these alien civilisations, that want to travel vast distances to buzz a passenger airplane or abduct seemingly normal, random people, would have the same constraints as us.

They would have the same problems we faced, firstly escaping their planet's gravity!

Well, we know from a previous chapter that it's not easy.

They would need to mine a fuel to do this, or certainly mine material to capture the energy from a source. Also, they would require metals, or a similar substance, hardy enough to withstand space travel.

These are not things that could be done overnight, it would take years, decades, centuries, millennia of evolution. Then to fabricate machines would mean probably generations of co-operative working and raising families. Raising families and co-operative working would indicate empathy and understanding.

In my opinion, to get this far, they would not be looking to be hostile, they would be looking to say, "Hey, look at us. You're not alone. We worked out how to say hi".

Of course, I could be totally wrong, and they may be led by a tyrant, squid-faced leader who is bent on ruling the universe or creating equilibrium like Thanos from the Marvel films. In fact, they may be on their way here now! It is, of course, possible.

I doubt it, though. The trouble is when you watch these movies, they make space travel seem easy; just whack it into warp drive and, after a very short time, you arrive at your destination.

Unfortunately, this isn't the case. Both Newton and Einstein showed us that physics is the same wherever you are in the universe. The same natural laws apply here as they would on

Chapter 9 - Alien Civilisations

any other planet in this solar system or planet in another galaxy many millions of light years away.

When you get to realise just how staggeringly vast space is, you begin to understand the reality of ever visiting another planet outside our own solar system.

The Voyager probes will come close to another star system in around 40,000 years, but will still be about another 40,000 years away from an actual star. Remember, they are travelling at around 35,000 miles an hour. That's 80,000 years! More than ten times longer than we have history for our species.

OK, let's say with technological advances we can double this speed to 70,000 miles per hour, it's still a distance that would take 40,000 years. That means we would be relying on generation after generation successfully breeding, which, by the way, we don't know if it is even possible in space. Let's not also forget the children! Generations of curious offspring constantly asking: "Are we there yet?"

It's not beyond the realms of possibility that we could sustain a food source, as water is relatively easy to make when you know the formula, and, possibly, we could make artificial sunlight.

The problem would be the human factor; one generation not reproducing, which is highly probable due to being in space where the body ceases to function properly.

Hey, let's assume, for this chapter's sake, we could get over that.

How about the mental state of these traveller's though?

What would happen if they were born with disabilities or were mentally impaired?

OK, let's, just for the heck of it, presume we managed to create an ark-type spacecraft, travelling at several hundreds of thousands of miles per hour, with thousands of people on board so there was a decent gene pool and so people didn't have to reproduce with their relatives. Also, food and water were no problem, as we had managed to create artificial light. Somehow, every generation was successful and giving birth was no problem.

In 10,000 years, we got to our closest star, Proxima Centauri, and found a planet. Then what?

The chances of a planet being suitable for life as we know it are staggeringly low. It could be a gas planet, or an ice planet, or it could be a furnace. Its atmosphere could be toxic to us; the atmosphere on every other planet we have found so far is toxic. Would the star it orbits be a stable star?

It's so incredibly, microscopically unlikely that the first, or closest, planet we get to could or would be suitable for human existence.

So then what? Well, I guess it's back in the ark for another trip, which will take tens of thousands of years, to reach the next star system.

It really doesn't seem at all possible.

Of course, there may be other ways; wormholes that bring

you out to another part of the universe, for instance. These are theoretical but definitely possible. However, first you would have to find one (I'm inclined to think this is easier said than done), and then you don't know where it's going to cosmically spit you out or eject you. The chances are it will be tens of thousands of years from anywhere. Space is a really big place and mostly empty. In fact, you could say it's relatively spacious!

I'm sorry, but for me, sending people into deep space is a definite no go. We may someday manage to send manned missions out to the gas and ice giants of our own solar system, but I think that will be about it.

This isn't to say, of course, that there isn't life out there that could be contacted or that could contact us.

Enrico Fermi was a brilliant scientist. Born in Rome in 1901, his work laid the groundwork for the US to make the atomic bombs that all but ended World War II.

He was also a Nobel prizewinner for his work on discovering new radioactive elements.

So, when Fermi was having lunch one day and he stated that if there was intelligent life out there where was it? People took notice.

Where is everybody? He realised that with the age of the galaxy, the Milky Way could have been overrun with intelligent life. But it wasn't, or isn't and, as far as we know, never has been.

It's called the Fermi paradox, which suggests there is no

intelligent life because, if there was, it would have contacted us by now.

Radio astronomer Frank Drake took it one more step, in 1961. He created an equation to work out the possibility of life in the Milky Way.

Here it is:

$$N = R^* f_p n_e f_l f_i f_c L$$

N: Stands for the number of civilisations in the Milky Way where communication might be possible

R^*: Stands for rate of star formation in the galaxy

f_p: The fraction of stars with planetary systems

n_e: The number of potential planets where life could possibly have started per star.

f_l: The fraction of planets where life develops

f_i: The fraction of planets where intelligent life evolves

f_c: The fraction of planets that become able to communicate with interstellar radio

L: The average lifetime of these civilisations

Drake and his team of scientists were very conservative with the figures, but it still came out at somewhere between ten

Chapter 9 - Alien Civilisations

thousand and a hundred million civilisations that should be out there. Back to the Fermi paradox! "Where is everybody?"

I am never one to question scientists much more intelligent than me, but I have an issue with this equation. I mean, just how conservative were they with fi (the fraction of life that develops intelligent life)? I mean, for 3.5 billion years life has developed on this planet and, up until a few thousand years ago, we couldn't have called any life anywhere near capable or with the potential of interstellar communication.

I mean, just think how many species there have been before us and that there are now. It's in the billions, and the fossil record shows only a smidgen of what there was. There could have been billions of other species that didn't leave fossil records. It's kind of pot luck, a chance in a billion, to become a fossil, and with all that life, just one accidental freak of nature that created us.

Without us, imagine all that life, gorillas, whales, dogs, cats, lemurs, dragonflies, mice, and yet not one of them even capable of using, or let alone making, a spanner.

Despite all this, we are always on the lookout for life out there though, thanks to one incredible organisation: **SETI**

Its purpose? To search for extraterrestrial life.

That's what they do. They scan the night sky looking for signs of intelligent life, having probably realised that you can't just zip around the cosmos looking for life on planets, or expect life to find its way here. The best way to find evidence of extraterrestrial life is to rely on radio telescopes.

Large antennas listen out for any signals aliens maybe transmitting. Radio would be the best way to communicate, as it travels at the speed of light, making it possible to bridge the gap of enormous distances.

SETI might be listening, but it's not talking. SETI does not transmit any messages or data into space.

How did it all get started then?

Well, in 1959, a couple of physicists realised that a large enough transmitter could pick up data from a large receiver or antenna despite being separated by incredibly vast distances. Even interstellar expanses!

Philip Morrison and Giuseppe Cocconi realised that they had the ability to detect signals from another world or civilisation who maybe sending signals!

The first person to actually probe the stars for a signal was our friend Frank Drake (responsible for the Drake equation we mentioned earlier). In 1960, he used an enormous antenna in Green Bank, West Virginia pointed towards two large star systems.

This was unsuccessful but is generally accepted as the start of SETI.

In the late 1980s, NASA funded SETI, but by the early nineties the funding was pulled by the American government, and SETI had to seek funding from somewhere else.

Chapter 9 - Alien Civilisations

Funds now come from private investors and the general public including, albeit in a very small way, me! Anyone can help the SETI fund. Look at the website for more information.

Anyway, thanks to these private funds, SETI has been able to keep its eyes and ears open to the cosmos.

Life appeared relatively quickly on earth after it had formed, so it would be reasonable to think that this is an almost certain outcome if the conditions were similar to earths, or at least where liquid water was present. Ever wondered why they are always looking for signs of water on Mars?

NASA's Kepler telescope and the Hubble telescope have found plenty of such planets!

What would happen if we did receive a signal from another planet?

The SETI Institute says that the signal will reveal a few things about where it's coming from and about the planet from where it originated.

So then what?

Does that then mean an end to all religions?

Do paranoid government militaries start preparing for an imminent attack, or retaliation, despite how friendly the message may appear?

Would they even tell us? How would we cope knowing we are

not the most intelligent thing in the universe that we haven't made up or imagined?

SETI suspects that this would be quite a drawn-out communication though, as it can take years for signals to be sent, received, deciphered and then replied to. In the very unlikely event that it came from our closest neighbour, it would take over eight years for a response.

"Once an artificial signal is confirmed as being of extraterrestrial, intelligent origin, the discovery will be announced as quickly and as widely as possible," says SETI.

"There will be no secrecy, and indeed getting the word out quickly is important, as there would be an urgent need to have astronomers world-wide monitor any detecting signals 24 hours a day."

Hmm, I wonder what our governments would have to say about that?

According to SETI Institute astronomer Seth Shostak, there are three ways to find life on other worlds.

The first is to send a rocket or probe and see for ourselves. Not really possible, at this moment in time, unless we are looking in our own solar system.

The second is to use advanced telescopes like Hubble or the James Webb telescope. These advanced machines can detect light which will indicate what sort of atmosphere the planet has.

Chapter 9 - Alien Civilisations

Lastly is to monitor the skies, looking for any signs that could be construed as a signal or evidence of intelligence.

Oh, and of course, if we do ever find a signal, whether meant for us or one that we manage to come across that's maybe meant for someone else, will we ever be able to decode it?

Well, this is what SETI does, and it's a great institution. Anyone can sign up and help scan the skies for messages or signals. Visit their website if you are inclined in any way to do so.

Picture courtesy of SETI.ORG

Although SETI has had no luck so far, in 1977 we did experience the Wow! signal.

In 1977, at the Big Ear radio observatory, Ohio State University in Delaware, Ohio, a volunteer astronomer called Jerry Ehman was scrolling through data printouts from a previous day, when he noticed something rather extraordinary, a rather strong signal that lasted for exactly 72 seconds. He excitedly circled the anomaly in red pen and wrote the word "Wow!" next to it.

Relatively Spacious

Of course, to me and you these random numbers and letters don't mean anything but to Jerry, he realised immediately how strange this was. At the time, it was considered very strong evidence of extraterrestrial intelligence. The dilemma we have with it, though, is that it only appeared once and nothing slightly or remotely similar to it has been discovered since, which in some ways makes it even more intriguing.

Technology is far superior today compared to the seventies and, despite looking in this region again, we still have not found anything similar.

The detection was exciting though, and, in many ways, still is, especially when you consider this is exactly what SETI was looking for.

So, was it a message from an alien race or not?

"Nobody knows" Says SETI's Shostak.

The Wow! signal (highlighted in red by Jerry Ehman): picture courtesy of space.com

Chapter 9 - Alien Civilisations

Over the years, there have been a few theories. One suggests it was a comet nearing the sun, but that theory hasn't really held up too well. Scientists, I'm sure, will eventually figure it out; whether it was intelligent, or just a gamma burst from some exploding star, or just a cosmic burp from deep space.

So, are we alone in the universe or not? I would say probably not, but ever making contact with another civilisation, I think, is very, very doubtful.

So, in effect, we are on our own. We have to consider that there could well be life on other planets. Maybe they are like most of the creatures on earth, beautiful but not capable of interstellar communication. Or perhaps they didn't evolve to have eyes or ears. Life didn't evolve those features for billions of years here. The eye and the ear are, to all intents and purposes, a fortunate gene malfunction.

Maybe we should consider that there were civilisations out there, but they died out billions or millions of years ago. Or perhaps there are civilisations, but they are microscopic and their equipment is just not detectable by us.

Of course, there could be another reason we haven't heard from anybody else out there!

Maybe they have been watching us for thousands of years and realised what a bunch of murderous, belligerent fanatics we are.

They may have witnessed the Christian Crusades or, in fact, any religious wars where people were murdered for not believing in what the others believed in.

Or perhaps they have seen people prepared to attach themselves to high explosives to blow themselves, and innocent people, to pieces including children.

Yes, humans will and indeed have killed over pretty much anything.

Could we blame another race for ignoring us or hiding from us? Any creature or habitat that has come into contact with human beings has found that it has not ended well for them and, in most cases, results in extinction.

In vanity, we might think another race would want to make contact. In reality, it would probably be a bad idea for them.

We have to make the best of life that we can, and that means looking after our own amazing and beautiful little planet.

Earth maybe insignificant, in the grand scheme of things, but it is important to us, and we need to treat it that way, whatever it takes.

I want my children's children to live in a world where there are rhinos and tigers and polar bears, and in a world that teaches about the stars and planets and the universe.

Perhaps we can't reach the stars, but perhaps we don't need to. Sometimes, just to look, see and admire is enough.

The universe is relatively spacious. It is the highest of honours

Chapter 9 - Alien Civilisations

to exist in one of its celestial bodies. As far as we know, it is only us. In fact, as far as we know, it has only ever been us and maybe it will only ever have been us. The fundamental miracle of living, the ultimate gift of life, which has made it possible for us to observe, comprehend and study a universe that still holds so may tantalising secrets.

Bibliography / References / Resources

Al-Khalili, J. (2020). The World According to Physics. New Jersey: Princeton University Press.

Bell, J. (2016). The Interstellar Age: Inside the forty-year Voyager mission. New York, NY: Dutton.

Bryson, B., & Matthews, R. (2003). A Short History of Nearly Everything. Santa Ana, CA: Books on Tape.

Bryson, B., & Turney, J. (2019). Seeing Further: The Story of Science and the Royal Society. London: William Collins.

Chown, M. (2018). The Ascent of Gravity: The quest to understand the force that explains everything. New York: Pegasus

Chown, M., & Woodroffe, D. (2018). Infinity In The Palm of Your Hand. London: Michael Omara Books Limited.

Cohen, A., & Cox, B. (2019). The Planets. London: Harper Collins Publishers.

Clifton, T. (2017). Gravity: A very short introduction. Oxford: Oxford University Press.

Bibliography / References / Resources

Cox, B., & Forshaw, J. R. (2010). Why Does E=mc2?: (And Why Should We Care?). Cambridge, MA: Da Capo Press.

Dartnell, L. (2020). Origins: How the Earth Shaped Human History. London: Penguin Random House UK.

deGrasse Tyson, N. (2017). Neil deGrasse Tyson's Astrophysics for People in a Hurry (1st ed.). New York: W.W. Norton and Company Inc.

Deutsch, D. (2012). The Beginning of Infinity. New York: Penguin Books.

Dunkley, J. (2019). Our Universe: An Astronomer's Guide (1st ed.). Massachusetts: The Belknap Press of Harvard University Press Cambridge.

Dunlop, S., & Tirion, W. (2020). 2021 Guide to the Night Sky. Glasgow: Harper Collins Publishers.

Editors, C. R., (2016). The Space Shuttle Challenger Disaster: The history and legacy of NASA's most controversial tragedy. United States: CreateSpace Independent Publishing Platform.

Fernandez-Armesto, F. (2020). Out of Our Minds. London: UNIV OF CALIFORNIA Press.

Golub, L., & Pasachoff, J. (2014). Nearest Star. Cambridge: Cambridge Univ. Press.

Gribbin, J. (2020). Seven Pillars of Science. London: Icon Books Ltd.

Hawking, S., Redmayne, E., Thorne, K., & Hawking, L. (2018). Brief Answers to the Big Questions. London: John Murray Publishers.

Krauss, L. M. (2013). A Universe from Nothing: Why there is something rather than nothing. New York: Atria Paperback.

Melcher, E. (2021). The Extraterrestrial Big Hello. Independently published.

Rovelli, C. (2017). Reality Is Not What It Seems: The Journey To Quantum Gravity. Milton Keynes: Penguin Random House.

Shayler, D., & Harland, D. M. (2016). The Hubble Space Telescope: From concept to success. New York: Springer.

Stannard, R. (2008). Relativity: A Very Short Introduction. Oxford: Oxford University Press.

West, D. (2015). A Short Biography of the Astronomer Edwin Hubble. United States: CreateSpace Independent Publishing Platform

https://www.explainthatstuff.com/light.html

https://www.space.com/18647-how-big-is-mercury.html

https://www.nationalgeographic.org/encyclopedia/atmosphere/

https://www.dummies.com/education/science/physics/einsteins-special-relativity/

Bibliography / References / Resources

https://www.sciencemag.org/news/2019/10/some-universes-heavier-elements-are-created-neutron-star-collisions

https://www.npr.org/sections/thetwo-way/2017/12/20/572252060/watch-what-happens-when-two-neutron-stars-collide?t=1614519972440

https://www.wired.co.uk/article/what-happens-when-two-neutron-stars-collide

https://www.ligo.caltech.edu/system/media_files/binaries/386/original/LIGOHistory.pdf

https://physicsworld.com/a/how-does-ligo-detect-gravitational-waves/

https://physicsworld.com/a/catching-gravity-rolling-by/

https://www.space.com/31913-how-scientists-detected-gravitational-waves-ligo.html

https://www.newscientist.com/article/2100834-how-big-is-a-proton-no-one-knows-exactly-and-thats-a-problem/

https://www.britannica.com/place/Neptune-planet/Neptunes-discovery

https://theplanets.org/what-color-is-venus/

https://phys.org/news/2019-06-earth-heavy-metals-result-supernova.html

https://www.mtwilson.edu

https://www.universetoday.com/21999/10-interesting-facts-about-neptune/

https://solarsystem.nasa.gov/moons/overview/

https://www.nasa.gov/mission_pages/apollo/apollo11.html

https://www.space.com/16452-jupiters-moons.html

https://theplanets.org/callisto/

https://science.nasa.gov/science-news/science-at-nasa/2002/05apr_hitchhiker

https://space.stackexchange.com/questions/2076/how-far-do-you-have-to-be-from-earth-to-be-in-space

https://earth.stanford.edu/news/what-caused-earths-biggest-mass-extinction#gs.vcxxre

http://www.bbc.co.uk/earth/story/20160610-it-took-centuries-but-we-now-know-the-size-of-the-universe

https://lco.global/spacebook/stars/supernova/

https://britastro.org/node/13975

https://nineplanets.org/uy-scuti/

Bibliography / References / Resources

https://umarscience9spaceproject.weebly.com/the-end--uyscuti--largest-star-known-to-man.html

https://wwf.panda.org/discover/our_focus/forests_practice/deforestation_fronts_/

https://www.britannica.com/place/Amazon-Rainforest

https://www.nationalgeographic.com/environment/article/rain-forests

https://sci.esa.int/web/education/-/45751-the-huygens-probe-lands-on-titan

https://www.nasa.gov/mission_pages/station/research/news/15_ways_iss_benefits_eart

https://www.google.com/search?client=safari&rls=en&q=jhow+far+is+Jupiter+form+the+sun&ie=UTF-8&oe=UTF-8

https://www.rmg.co.uk/stories/topics/interesting-facts-about-venus

https://www.nature.com/articles/d41586-020-03258-5

https://www.universetoday.com/110276/why-the-asteroid-belt-doesnt-threaten-spacecraft/

https://www.nasa.gov/mission_pages/juno/main/index.html

https://www.seti.org/seti-research

https://astronomy.com/news/2020/09/the-wow-signal-an-alien-missed-connection

https://www.nobelprize.org/prizes/uncategorized/the-age-of-the-sun-2/

https://penningtonplanetarium.wordpress.com/2015/06/26/meet-the-largest-star-ever-discovered/

https://www.space.com/23135-diamond-rain-jupiter-saturn.html

https://earthsky.org/space/what-will-happen-when-our-sun-dies

https://edition.cnn.com/2020/11/27/world/earth-supermassive-black-hole-milky-way-scn/index.html

https://theplanets.org/how-long-would-it-take-to-get-to-jupiter/

https://solarsystem.nasa.gov/resources/2388/europa-3d-model/

https://www.greekmythology.com/Myths/The_Myths/Zeus%27s_Lovers/Ganymede/ganymede.html

https://www.ligo.org/science/GW-Multiple.php

https://nssdc.gsfc.nasa.gov/nmc/spacecraft/display.action?id=1971-051A

Bibliography / References / Resources

https://spie.org/publications/fg08_p02_spectrometerspectroscopespectrograph?SSO=1

https://history.aip.org/exhibits/cosmology/tools/tools-spectroscopy.htm

https://nasa.tumblr.com/post/156674617514/10-space-football-facts-you-probably-didnt-know

https://www.forbes.com/sites/scottsnowden/2019/03/12/solar-power-stations-in-space-could-supply-the-world-with-limitless-energy/

https://www.mccc.edu/~dornemam/Planet_Walk/Sun/the_sun.htm

https://www.space.com/18923-neptune-distance.html

https://www.bbc.co.uk/bitesize/guides/z3tb8mn/revision/3

https://www.space.com/38982-no-big-bang-bouncing-cosmology-theory.html

https://spacemath.gsfc.nasa.gov/weekly/10Page3.pdf

http://www.bradshawfoundation.com/privacy_policy.php

Gary Bridges born and raised in Tunbridge Wells is a keen amateur astronomer.

His lifelong passion for science has prompted him to develop the *Relatively Speaking* series.

Relatively Spacious, the first of the three books, about time and space.

Relatively Unstable, a book about the destructive forces on Earth

Relatively Insatiable, a book of the history of food and how we came to eat what we do.

Printed in Great Britain
by Amazon